Optimal Design

Editor:
Heinz Holling

Zeitschrift für Psychologie

Vol. 221, No. 3, 2013

HOGREFE

Library of Congress Cataloging in Publication

is available via the Library of Congress Marc Database under the
LC Control Number 2013944873

Library and Archives Canada Cataloguing in Publication

Optimal design / editor: Heinz Holling.

Published as part of the journal Zeitschrift für Psychologie, vol. 221, no. 3, 2013.
Includes bibliographical references.
ISBN 978-0-88937-448-5 (pbk.)

1. Psychology--Experiments. 2. Optimal designs (Statistics).
I. Holling, Heinz, 1950-, editor of compilation. II. Title: Zeitschrift für Psychologie

BF198.7.O68 2013 150.72'4 C2013-904705-0

PUBLISHING OFFICES
USA: Hogrefe Publishing, 38 Chauncy Street, Suite 1002, Boston, MA 02111
Phone (866) 823-4726, Fax (617) 354-6875, E-mail customerservice@hogrefe-publishing.com
EUROPE: Hogrefe Publishing, Merkelstr. 3, 37085 Göttingen, Germany
Phone +49 551 99950-0, Fax +49 551 99950-425, E-mail publishing@hogrefe.com

SALES & DISTRIBUTION
USA: Hogrefe Publishing, Customer Services Department, 30 Amberwood Parkway, Ashland,
OH 44805, Phone (800) 228-3749, Fax (419) 281-6883, E-mail customerservice@hogrefe.com
EUROPE: Hogrefe Publishing, Merkelstr. 3, 37085 Göttingen, Germany
Phone +49 551 99950-0, Fax +49 551 99950-425, E-mail publishing@hogrefe.com

OTHER OFFICES
CANADA: Hogrefe Publishing, 660 Eglinton Ave. East, Suite 119-514, Toronto, Ontario M4G 2K2
SWITZERLAND: Hogrefe Publishing, Länggass-Strasse 76, CH-3000 Bern 9

Hogrefe Publishing
Incorporated and registered in the Commonwealth of Massachusetts, USA, and in Göttingen, Lower Saxony, Germany

ISBN 978-0-88937-448-5

Contents

Editorial

Current Issues in Optimal Design

Heinz Holling

Institute of Psychology, University of Münster, Germany

The quality of psychological studies substantially depends on their experimental design. The design has to be specified before starting data collection. Wrong decisions concerning the design can only be corrected with great difficulty later on, if at all. Thus, considerable attention should be devoted to the experimental design.

An experimental design concerns all steps of how to perform the experiment. Therefore, many questions have to be answered, for instance: How many factors, levels per factor, and time points need to be investigated? Should the factors be fixed or random? How many experimental units should be selected for each condition? Which covariates have to be included?

Many criteria to evaluate the goodness of designs have been developed. Designs should provide precise estimates of parameters and allow for sound testing of hypotheses. Thus, small standard errors of the estimates and a high power of the test are preferable. Furthermore, internal and external validity are among other standards which are mainly affected by the experimental design.

Optimal design considers an important issue in experimental design. It specifies the distribution of the values of the independent variables, depending on statistical criteria determined a priori. When comparing the outcomes of two groups by a *t*-test, for example, a desirable criterion has to be specified first. Obviously, the difference of the means should be estimated as precisely as possible, resulting in a small standard error or variance of this difference. Based on this criterion and given a certain sample size, the proportion of subjects allocated to the experimental or, equivalently, to the control group has to be derived such that the criterion just outlined is best fulfilled.

Optimal design for simple regression implies determining the values of the independent variable and the number of observations at each selected value. For instance, in the mental rotation paradigm (Shepard & Metzler, 1971), the reaction time is analyzed depending on stimulus rotation. Here, the optimal design specifies the degrees of rotation to be selected as, for example, 0°, 45°, 90°, 135°, and 180°, as well as the frequency of stimulus presentation. Often, the stimuli with different degrees of rotation are presented with the same frequency. Whether such a design is a good decision depends on the statistical criterion to be optimized. In simple regression, several criteria may be specified, for example, minimizing the standard error of the intercept, the standard error of the slope, or the average standard error of both. Or the predicted reaction time for certain values of the independent variable should be as precise as possible.

Optimal design has a long history in statistics. It goes back to Smith (1918), who wrote the first explicit paper about an optimal design problem. Many important contributions to optimal design theory were made in the 1950s, for example, by Elfving (1952) and Kiefer and Wolfowitz (1959, 1960). The first monographs on optimal design were published by Fedorov (1972) and Silvey (1980).

In the beginning, optimal design for linear models was dominant. This area is now well developed (e.g., Schwabe, 1996). Optimal design for nonlinear models is more challenging since such designs in general depend on the parameters to be estimated. Several kinds of designs were developed to cope with this problem, such as adaptive, Bayesian, and maximin designs (Holling & Schwabe, in press). The development of such designs for nonlinear mixed models is one current topic of special research interest.

Although optimal design has a long history, it did not receive much attention in psychology for a long time. Several reasons may be responsible for this. First, experiments in psychology are often not as expensive as in other fields, such as engineering or chemistry – high effort and cost motivate to thoroughly plan experiments. Second, textbooks on experimental design usually do not include a chapter on optimal design. Third, much of the literature about optimal design is written rather technically and requires profound statistical knowledge. Finally, software packages frequently used in psychology hardly provide opportunities to generate optimal designs.

Recently, some more introductory books have been published. First, the monograph by Berger and Wong (2009) should be mentioned. This comprehensively presents most of the important topics of optimal design relevant for psychological research. Goos and Jones (2012) introduce optimal design using a case study approach. A more advanced, often cited monograph was written by Atkinson, Donev, and Tobias (2007). A comprehensive overview of optimal design theory which, however,

DOI: 10.1027/2151-2604/a000141

presupposes advanced statistical knowledge, is provided by Pukelsheim (2006). An introductory article emphasizing optimal design in psychology has been published by McClelland (1997), while Holling and Schwabe (in press) provide an overview of optimal design theory stressing issues of item response theory.

Application of statistical methods, in general, requires statistical software. Many popular statistical software packages offer few opportunities for generating optimal designs, if any. Exceptions are the software packages SAS and JMP from SAS Institute Inc., which include several procedures for generating optimal designs. There is a growing number of programs for optimal design written in R (e.g., Grömping, 2011; Rasch, Pilz, Verdooren, & Gebhardt, 2011). Furthermore, many stand-alone programs or programs written in Matlab or Mathematica are available for generating optimal designs for specific statistical tasks.

The demand for optimal designs in psychological research has been growing in recent years. Neuroimaging methods like fMRI have become a standard procedure in psychological research. Since these methods are rather complex and expensive, optimal design may considerably reduce the time and cost involved (Maus & van Breukelen, 2013). Optimal design is extremely valuable for large-scale assessments, for example, the Programme for International Student Assessment (PISA; OECD, 2012) or the Trends in International Mathematics and Science Study (TIMSS; Mullis, Martin, Ruddock, O'Sullivan, & Preuschoff, 2009). Such studies require large sample sizes, which may be kept to a minimum by using optimal design (see Kuhn & Kiefer, 2013). Adaptive testing, which minimizes the number of items, also requires optimal design (see Holling & Schwabe, in press). However, practically every psychological study may benefit from optimal design, since the reduction of observations necessary is often considerable.

This topical issue provides an introduction to optimal design and an overview on new developments and applications of optimal design to several important topics in psychology. These topics consist of multilevel analysis, event history analysis, functional magnetic resonance imaging (fMRI) studies, and optimal test assembly. All authors have paid intense attention to high readability. Thus, these articles only presuppose statistical knowledge as typically provided by the BSc and MSc study of psychology. More technical issues are always deferred to an appendix.

The first article by Holling and Schwabe (2013) provides an introduction to optimal design using some typical psychological research issues. The basics of this subject are presented by means of the *t*-test and simple regression, two common statistical procedures. Optimal designs are easily derived for both methods and it is shown how much additional sample size is required to achieve the same statistical performance when using nonoptimal designs. Then, optimal design is more generally outlined for linear models. Several optimality criteria that have been proposed in the literature are introduced and it is shown that the selection of an adequate criterion to be optimized depends on the hypotheses of a study. Developing optimal designs for nonlinear models is much more demanding than for linear models. Here, as a basic problem, the design to be created depends upon the unknown parameters. This issue is illustrated by logistic regression with a binary predictor. This simple nonlinear statistical method is also used to fundamentally explain adaptive, Bayesian, and minimax designs.

Multilevel studies are frequently encountered in psychological research. Experimental or observational units are frequently nested in groups or clusters, for example, pupils in school classes, employees in organizations, or repeated measurements within persons. Data have to be sampled at different levels – the unit level and the group level. Thus, optimal design for multilevel studies concerns the levels of the independent variables as well as the sampling of units at the different levels of the study. Since multilevel studies usually require large samples of subjects and clusters, optimal design is especially important for this kind of modeling. Van Breukelen (2013) gives a comprehensive and clearly written introduction to optimal design for multilevel studies. Optimal designs of cluster randomized trials are presented and illustrated by an example regarding stress management in primary school. Then, optimal designs of multisite trials are outlined as well as models that include covariates. Furthermore, the author discusses optimal designs for other important multilevel models such as models with dichotomous outcomes or nested experiments with two and three levels. Finally, hints for useful software are given.

In the contribution by Moerbeek and Jóźwiak (2013), optimal designs for event history analysis are derived. Event history analysis, also known as survival analysis, is a statistical method to analyze the occurrence and timing of events. Such events may be job loss, an accident at work, or premature psychotherapy termination. The occurrence and timing of events can be related to covariates and points in time with a high probability of the event occurrence being identified. Therefore, event analysis can be usefully applied in many psychological disciplines, such as clinical or work and organizational psychology. Moerbeek and Jóźwiak (2013) consider event history analysis with discrete time intervals. Here, it is not necessary to know the exact time points at which the events occur. Before deriving optimal designs, the authors give a comprehensible introduction to event history analysis. In optimal design for such models, the number of subjects and time intervals has to be determined. Moreover, the authors analyze the effects of costs and attrition on the optimal designs. Their results are illustrated by an example on drinking onset. Finally, the authors present a user-friendly computer program that facilitates the calculation of optimal design for event history analysis for researchers.

In recent years, functional magnetic resonance imaging (fMRI) has rapidly emerged as a versatile technique in psychology as well as in neuroscience. It is, however, a very elaborate and expensive method. For this very reason, optimal design provides a useful technique to increase the efficiency of fMRI experiments. Maus and van Breukelen (2013) present a comprehensive overview of optimal design for fMRI studies. The authors first introduce the fundamental methodology of such studies. They describe common objectives of two basic types of fMRI experiments, blocked and event-related designs, and outline statistical models used for data analysis of this kind of studies. Then, the most important results for optimal design of blocked and event-related designs concerning different design objectives are explained and it is shown that fMRI studies can considerably benefit from optimal design.

Kuhn and Kiefer (2013) describe the optimization procedures used in generating the design of a large-scale assessment of educational standards in mathematics in Austria. The main purpose of large-scale assessments is to quantify the degree of mastery that a population of interest has obtained in a specific content domain (e.g., mathematics). Parameters at the population level are of key interest here. However, the design of large-scale assessments is a complex endeavor, as several, sometimes competing goals have to be taken into consideration simultaneously (e.g., unbiased estimation of population parameters, longitudinal linking across assessment cycles). Kuhn and Kiefer (2013) describe the two-step procedure used in generating the test design of the educational standards assessment of mathematics in Austria. First, they used a local search heuristic, that is, simulated annealing, to generate a nearly optimal design at the item block level. Second, they used linear optimization to allocate items from a large item pool to the item blocks. In addition to providing an introduction to optimal test design in large-scale assessment, their work carefully outlines decisions and constraints made in the design process that are often not explicitly mentioned in the literature, thus allowing a more detailed grasp of the complexities faced in designing large-scale assessments.

Acknowledgment

This article and editing of the topical issue were partly supported by the German Research Foundation (DFG) under Grant No. HO 1286/6-2. I would like to thank all authors for providing the articles in this topical issue. I am especially grateful for their willingness to contribute comprehensible overview articles. In addition, many thanks go to the anonymous reviewers who gave constructive criticism and detailed feedback.

References

Atkinson, A. C., Donev, A. N., & Tobias, R. D. (2007). *Optimum Experimental Designs, with SAS*. Oxford, UK: Oxford University Press.

Berger, M. P. F., & Wong, W. K. (2009). *An introduction to optimal designs for social and biomedical research*. Chichester, UK: Wiley.

Elfving, G. (1952). Optimum allocation in linear regression theory. *Annals of Mathematical Statistics, 23*, 255–262.

Fedorov, V. V. (1972). *Theory of Optimal Experiments*. New York, NY: Academic Press.

Goos, P., & Jones, B. (2012). *Optimum designs of experiments: A case study approach*. Chichester, UK: Wiley.

Grömping, U. (2011). Tutorial for designing experiments using the R package RcmdrPlugin.DoE. *Reports in Mathematics, Physics and Chemistry*. Berlin: Department II, Beuth University of Applied Sciences.

Holling, H., & Schwabe, R. (2013). Introduction to optimal design: Some basic issues using examples from dyscalculia research. *Zeitschrift für Psychologie, 221*, 124–144.

Holling, H., & Schwabe, R. (in press). Optimal Design Theory. In W. J. van der Linden & R. K. Hambleton (Eds.), *Handbook of Item Response Theory* (Vol. 2). Boca Raton, FL: CRC Press.

Kiefer, J., & Wolfowitz, J. (1959). Optimum designs in regression problems. *Annals of Mathematical Statistics, 30*, 271–294.

Kiefer, J., & Wolfowitz, J. (1960). The equivalence of two extremum problems. *Canadian Journal of Mathematics, 12*, 363–366.

Kuhn, J.-T., & Kiefer, T. (2013). Optimal test-assembly in practice: The design of the Austrian educational standards assessment in mathematics. *Zeitschrift für Psychologie, 221*, 190–200.

Maus, B., & van Breukelen, G.J.P. (2013). Optimal design for functional magnetic resonance imaging experiments: Methodology, challenges and future perspectives. *Zeitschrift für Psychologie, 221*, 174–189.

Moerbeek, M., & Jóźwiak, K. (2013). Optimal designs for event history analysis. *Zeitschrift für Psychologie, 221*, 160–173.

McClelland, G. H. (1997). Optimal design in psychological research. *Psychological Methods, 2*, 3–19.

Mullis, I. V., Martin, M. O., Ruddock, G. J., O'Sullivan, C. Y., & Preuschoff, C. (2009). *TIMSS 2011 assessment frameworks*. Chestnut Hill, MA: TIMSS & PIRLS International Study Center.

OECD. (2012). *PISA 2009 technical report*. Paris, France: OECD.

Pukelsheim, F. (2006). *Optimal design of experiments*. Philadelphia, PA: SIAM.

Rasch, D., Pilz, J., Verdooren, R., & Gebhardt, A. (2011). *Optimal experimental design with R*. Boca Raton, FL: CRC Press.

Schwabe, R. (1996). *Optimum designs for multi-factor models*. New York, NY: Springer.

Shepard, R. N., & Metzler, J. (1971). Mental rotation of three-dimensional objects. *Science, 171*, 701–703.

Silvey, S. D. (1980). *Optimal design*. London, UK: Chapman & Hall.

Smith, K. (1918). On the standard deviations of adjusted and interpolated values of an observed polynomial function and its constants and the guidance they give towards a proper choice of the distribution of observations. *Biometrika, 12*, 1–85.

Van Breukelen, G. J. P. (2013). Optimal experimental design with nesting of persons in organisations. *Zeitschrift für Psychologie, 221*, 145–159.

Heinz Holling

Faculty of Psychology and Exercise and Sport Studies
Chair of Statistics and Methods
University of Münster
Fliednerstr. 21
48149 Münster
Germany
Tel. +49 251 83-38485
Fax +49 251 83-39469
E-mail holling@uni-muenster.de

Review Article

An Introduction to Optimal Design

Some Basic Issues Using Examples From Dyscalculia Research

Heinz Holling[1] and Rainer Schwabe[2]

[1]Institute of Psychology, University of Münster, Germany,
[2]Institute of Mathematical Stochastics, University of Magdeburg, Germany

Abstract. In this article we give an elementary introduction to optimal design for some basic statistical models. Research questions coming from a project on dyscalculia are used as illustrative examples throughout this article. First, basic design issues are considered for the *t*-test and simple regression and then extended to the general linear model. Finally, we will outline optimal designs for nonlinear statistical models. As a simple example, logistic regression with a binary explanatory variable is used.

Keywords: optimal design, *t*-test, simple regression, general linear model, nonlinear models

The outcome of an experiment essentially depends on the experimental conditions. The experimental design specifies the experimental conditions to be chosen and the number of replications for these conditions. Such a design has to be specified before any data have been collected. Afterwards, statistical analyses can usually not compensate for wrong decisions concerning design issues. Important design questions concern the kind of design to be selected, for example, factorial design, block design, or Latin square design, as well as sample sizes or to guarantee a prespecified power. These decisions depend on several conditions, such as the hypotheses, the circumstances, and budget of a study.

One basic task when determining a design is the assignment of respondents or, more general, experimental units to the values of the explanatory variables, also called independent variables or predictors. For example, if the impact of children's height on their body weight is to be analyzed by simple regression, different groups of children each with a specific body height can be selected for such a study. In the case of a two sample test, for example *t*-test, the independent variables are given by the membership to one of the two groups. A binary variable X with two levels, for example experimental group = 1, control group = 0, may define the group membership. For a 4^3 ANOVA, three variables (factors) are the independent ones, for example X_1, X_2, and X_3 with values (levels) 1, 2, 3, and 4 for each variable.

The main goal of optimal design is to specify the frequency distribution of the independent variables in order to get an optimal value of a specific statistical criterion, for example smallest standard error of parameter estimators. For an optimal design, first a statistical quality criterion has to be explicitly stated, then, the frequency distribution will be specified to meet the criterion as well as possible.

Even though optimal design theory has a long history in statistics, it has not received much attention within social sciences for a long time. Meanwhile, interest in optimal design is increasing rapidly as such designs are needed in large-scale assessment, for example PISA, or for adaptive testing. In general, optimal design may provide considerable advantages for nearly any statistical analysis, such as the *t*-test, simple and multiple regression, or item response models.

A very simple example is the proportion of subjects assigned to the groups of a *t*-test in order to get a small confidence interval for the estimated difference of the means of the dependent variable between both groups and accordingly a high power for the *t*-test. This will be the first example we will consider after providing a short outline of the illustrative example used in this article.

Effectiveness of Training for Children With Dyscalculia

Dyscalculia comprises severe difficulties in understanding and correctly applying calculation and arithmetical concepts. More fundamental numerical capabilities like counting, magnitude comparison, or number line estimation are usually affected as well (e.g., Landerl, Bevan, & Butterworth, 2004). For example, a recent study showed

DOI: 10.1027/2151-2604/a000142

Independent Samples Test

		Levene's Test for Equality of Variances		t-test for Equality of Means				
		F	Sig.	t	df	Sig. (2-tailed)	Mean Difference	Std. Error Difference
Difference of addition scores	Equal variances assumed	.157	.697	-3.009	18	.008	-3.70000	1.22972
	Equal variances not assumed			-3.009	16.575	.008	-3.70000	1.22972

Figure 1. Results from a *t*-test for the gain scores of the subtest addition in the prestudy.

that 10-year-old children with dyscalculia are comparable to 5-year-old control children in their ability to discriminate magnitudes (comparing two sets of dots; Piazza et al., 2010). In an ongoing study (Kuhn, Raddatz, Holling, & Dobel, 2013), a computer-based training for children with dyscalculia – or, more precisely, developmental dyscalculia – is being developed and its effectiveness evaluated. The current international classification systems ICD-10 (World Health Organization [WHO], 2010) and DSM-IV-TR (American Psychiatric Association [APA], 2000) show a large conceptual overlap in their criteria for defining dyscalculia: Firstly, a child's score in a standardized mathematics assessment should fall below a prespecified percentile rank cutoff (usually 10%); secondly, intelligence is required to be in the normal range (i.e., IQ $\geq$ 70). And thirdly, a substantial discrepancy between the mathematics and intelligence standard scores should be observed. In this paper, we will define children that conform to all three criteria as dyscalculic, whereas children who fulfill only the first two criteria will be defined as having a mathematical learning disability (MLD). The training is developed for children with dyscalculia and mathematical learning disabilities, respectively. Several design questions arise, some of which are used here to elucidate the impact of optimal design.

One of the central research questions within this project relates to the effectiveness of two newly developed training variants. The first one focuses on numerical precursor skills exclusively (as described above), while the other one refers to both numerical precursor skills and working memory (especially central executive tasks with spatial content, as this capacity is often impaired in children with dyscalculia (Schuchardt, Maehler, & Hasselhorn, 2008). These elaborate trainings stretch over 30 days and are individually adapted such that accuracy rates vary between 75% and 90% of training tasks solved correctly. To measure training effectiveness, the Heidelberger Rechentest (HRT 1-4), a standardized test of arithmetic skills (Haffner, Baro, Parzer, & Resch, 2005), is used. Two further groups serve as control groups. The first control group receives a computer-based training of inductive reasoning (Lenhard, Lenhard, & Klauer, 2012) such that a comparison between more domain-specific trainings and a domain-general training is possible. Finally, a passive waiting list control group does not receive any intervention at all in order to serve as a baseline for comparisons. Thus, in total, the current evaluation study comprises two training groups and two control groups. A lot of more specific questions have been posed, which will be taken up when discussing optimal designs for certain statistical models.

Optimal Design for *t*-Test

In a prestudy, which will be used as the first example here, the effectiveness of a first version of the training focusing solely on numerical precursor skills was investigated. Arithmetic skills have been measured by the HRT 1-4 before and after the training. The difference, that is *test score after the test – test score before the test*, represents the gain of arithmetic skills and is to be computed for the members of the training as well as a passive waiting control group. The training is deemed successful if the mean difference scores are significantly higher for the training group than for the control group. The budget allowed for a sample of n = 20 children altogether.

The question concerning optimal design is: How many out of the 20 children should be allocated to the training or experimental group? When answering this question we will refer to the results of an empirical study where 10 children were assigned to each group.

Assuming homogeneous variances for both populations (see Figure 1), a *t*-test for independent samples was run by SPSS. The results of this *t*-test for the scores of the subtest *addition* of the HRT 1-4 are shown in Figure 1. (For the other subtests similar results were obtained.) As the results indicate, a significant difference in the increase of the addition scores occurred. Thus, the results support the hypothesis that the training is a successful intervention.

The scores for the experimental and control group will be denoted by Y_E and Y_C and the corresponding sample sizes by n_E and n_C, respectively. Then, the effectiveness of the training is evaluated by the difference $\delta = \mu_E - \mu_C$, where μ_E and μ_C are the means of the population of trained and not trained children, respectively. δ is estimated by the difference of the corresponding sampling means, that is $\hat{\delta} = \hat{Y}_E - \hat{Y}_C$.

The success criterion δ, measured by several subtests of the HRT 1-4 is assumed to be independently identically distributed according to a normal distribution with constant variance σ^2 for the experimental and control group. The test statistic of the *t*-test, which provides an adequate statistical procedure, consists of the estimated difference of the corresponding sampling means, $\hat{\delta} = \hat{Y}_E - \hat{Y}_C$, divided by its standard error $SE(\hat{\delta})$. This standard error is given by

$$SE\left(\hat{\delta}\right) = \sqrt{\sigma^2\left(\frac{1}{n_E} + \frac{1}{n_C}\right)}. \quad (1)$$

Usually the variance σ^2 is unknown. It will be estimated by

$$\hat{\sigma}^2 = \frac{(n_E - 1)S^2_{Y_E} + (n_C - 1)S^2_{Y_C}}{n_E + n_C - 2}, \quad (2)$$

where $S^2_{Y_E}$ and $S^2_{Y_C}$ are the sample variances of Y for the experimental and the control group, respectively. This estimated variance will then be used for computing the confidence interval and test statistic. The optimal design does not depend on the values of the variance σ^2. Thus, for the sake of simplicity, we assume the variance σ^2 to be known with $\sigma^2 = 1$. As we will show next, $\mathrm{Var}(\hat{\delta})$ depends on the proportion of subjects assigned to the experimental group (or equivalently to the control group).

To get a small variance of $\hat{\delta}$ for a given $n = n_E + n_C$, we only have to minimize $\frac{1}{n_E} + \frac{1}{n_C}$, since σ^2 is a constant and cannot be influenced by the experimenter. Because of $\frac{1}{n_E} + \frac{1}{n_C} = \frac{n}{n_E n_C}$ we will minimize $\frac{n}{n_E n_C}$, or equivalently, maximize $n_E\, n_C$.

It is easy to show for an even n that $n_E\, n_C$ is maximal or, equivalently, $\mathrm{Var}(\hat{\delta})$ is minimal when $n_E = n_C = n/2$ (see Figure 1). If n is odd, no unique solution exists, the minimal variance of $\hat{\delta}$ is given when $n_E = n/2 - 1/2$ and $n_C = n/2 + 1/2$ or when $n_E = n/2 + 1/2$ and $n_C = n/2 - 1/2$ (see Figure 2). Thus, for $n = 21$ the optimal design is given by the sample sizes $n_E = 10$ and $n_C = 11$, or $n_E = 11$ and $n_C = 10$.

We can embed the above optimization problem for discrete values of the sample sizes into a continuous problem by using real numbers w_E and w_C for the proportions instead of $\frac{n_E}{n}$ and $\frac{n_C}{n}$, respectively, with $0 \leq w_E, w_C \leq 1$, and $w_E + w_C = 1$. Design problems based on real weights are in general easier to solve.

For the above situation, we have to maximize the real function $h(w_E) = w_E\, w_C = w_E\,(1 - w_E)$. Equivalently, the real function $g(w_E) = -w_E\,(1 - w_E)$ with $0 \leq w_E, w_C \leq 1$ can be minimized. The function $g(w_E)$ is convex and, thus, has a unique minimum. Since $g'\left(w^*_E\right) = 2w^*_E - 1 = 0$, its minimum is given for $w^*_E = 1/2$ (see Figure 3). Therefore, the proportion of the sample size of both groups, the experimental and the control group, should be .5 in accordance with the findings for optimal n_E.

Designs can be described in two ways. First, the optimal design for the prestudy referred to above is given by the proportions $\frac{n_E}{n} = 10/20$ and $\frac{n_C}{n} = 10/20$, Here, n_E, n_C, and n are integers. Thus, the designs described by proportions are called *exact*. Exact designs can be realized in applications. Second, we can describe designs by weights w_E and w_C. However, the products $w^*_E n$ and $w^*_C n$ will not necessarily yield an integer value and have to be rounded for determining the group sizes. Thus, designs defined by real weights are called *approximate* designs. Approximate designs have the advantage that they can be determined independently of the sample size.

More formally, exact and approximate designs are defined as follows. Both designs differ in the kind of the distribution of the design points. The set of all possible design points x is called the design region and denoted by X. Let the design consist of m distinct design points x_i, which occur n_i times in the experiment, a total number of $n = \sum_i^m n_i$ trials is given.

An exact design is defined by

$$\xi_n = \begin{Bmatrix} x_1 & x_2 & \cdots & x_m \\ \frac{n_1}{n} & \frac{n_2}{n} & \cdots & \frac{n_m}{n} \end{Bmatrix}. \quad (3)$$

Here, the distribution of the design points corresponds to a relative frequency distribution.

Optimal designs will usually be more easily generated when the denominator in weights n_i/n is not restricted to integer values. Thus, an approximate design ξ is defined by

$$\xi = \begin{Bmatrix} x_1 & x_2 & \cdots & x_m \\ w_1 & w_2 & \cdots & w_m \end{Bmatrix} \quad (4)$$

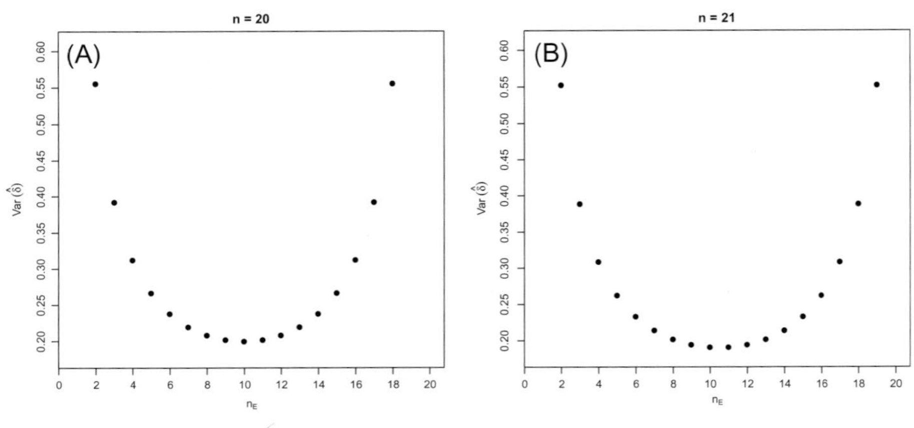

Figure 2. Impact of n_E on $\mathrm{Var}(\hat{\delta})$ for $n = 20$ (A) and $n = 21$ (B).

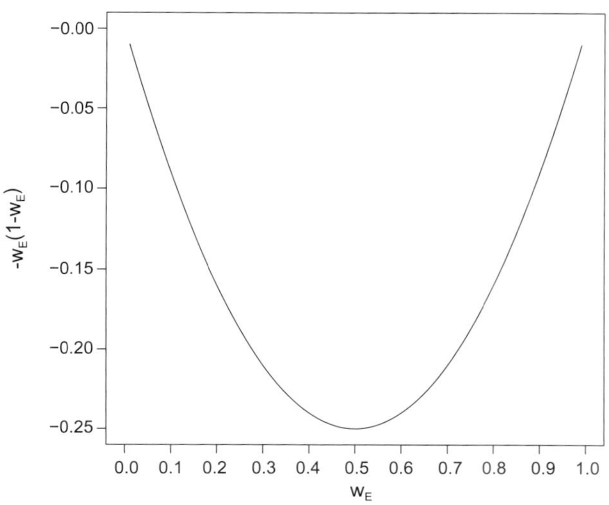

Figure 3. Graph of the function $g(w_E) = -w_E(1 - w_E)$.

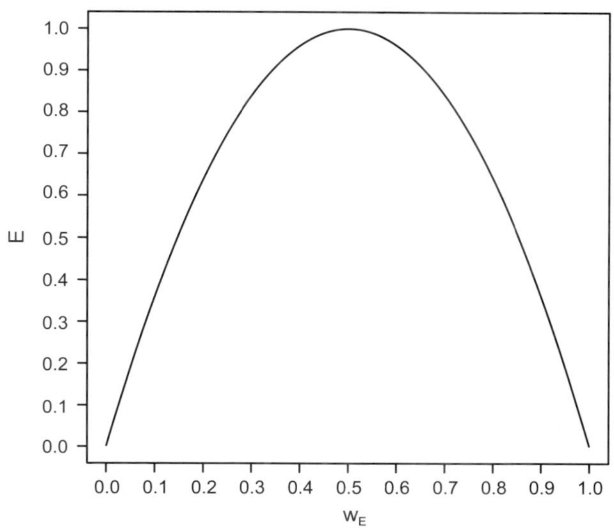

Figure 4. Graph of the efficiency for approximate designs of the prestudy.

with real-valued weights with $0 \leq w_i \leq 1$, $i = 1,\ldots, m$, and $\sum_{i=1}^{m} w_i = 1$. The distribution may be conceived of as a probability distribution.

For the prestudy the exact optimal design is given by

$$\xi_{10} = \begin{Bmatrix} 0 & 1 \\ 10/20 & 10/20 \end{Bmatrix} \tag{5}$$

and the corresponding approximate design by

$$\xi = \begin{Bmatrix} 0 & 1 \\ .5 & .5 \end{Bmatrix}. \tag{6}$$

We will now define two related measures for a design, the efficiency and deficiency, to describe the quality of a design. The minimum of $\mathrm{Var}(\hat{\delta})$, that is $\mathrm{Var}(\hat{\delta})$ for the optimal design and denoted by $\mathrm{Var}(\hat{\delta}^*)$, is given for our prestudy with $n = 20$ as

$$\mathrm{Var}\left(\hat{\delta}^*\right) = \sigma^2\left(\frac{1}{\frac{n}{2}} + \frac{1}{\frac{n}{2}}\right) = \sigma^2\,\frac{4}{n}. \tag{7}$$

If we relate $\mathrm{Var}(\hat{\delta}^*)$ to $\mathrm{Var}(\hat{\delta})$ of any design, we receive the *efficiency*

$$E = \frac{\mathrm{Var}\left(\hat{\delta}^*\right)}{\mathrm{Var}\left(\hat{\delta}\right)}. \tag{8}$$

The efficiency is a measure of the quality of a certain design compared to the optimal design. For the prestudy the efficiency, denoted by E, is given by

$$E = \frac{\sigma^2\,\frac{4}{n}}{\sigma^2\left(\frac{1}{n_E} + \frac{1}{n_C}\right)} = \frac{4 n_E n_C}{n^2} = 4 w_E\,(1 - w_E), \tag{9}$$

and depicted in Figure 4.

A further more intuitive measure based on the efficiency is the *deficiency* which is well suited for practical purposes. Deficiency measures the additional sample size of a design such that it is as efficient as the optimal design.

Now we will determine the factor $f = \frac{n}{n^*}$, which is the ratio of the sample size n of a certain design to the sample size of the optimal design n^*, such that $f\,\mathrm{Var}(\hat{\delta}^*) = \mathrm{Var}(\hat{\delta})$. Therefore, it follows that $f = \frac{\mathrm{Var}(\hat{\delta})}{\mathrm{Var}(\hat{\delta}^*)} = E^{-1}$. If, for example $f = 2$, it means that the sample size of a design has to be twice as large as that of the optimal design to get the same variance or equivalently the same standard error of the statistic $\hat{\delta}$. Now, we define the deficiency D as

$$D = E^{-1} - 1. \tag{10}$$

D reflects the ratio of the sample size for a certain design that has to be investigated additionally in order to achieve the same variance of the statistic $\hat{\delta}$ provided by the optimal design. (D* 100) % corresponds to the percentage of the additional sample size of a design such that it is as efficient as the optimal design. This measure is given in Figure 5 for approximate designs of the prestudy.

The above example, where the training success is evaluated by using a t-test, illustrates the role of optimal design within an empirical study. In a first step, the research question has to be specified. In the above example the training itself, the measurement scales, the population to be analyzed, and other issues have to be determined. Second, the design has to be specified. In our case, a pretest posttest design with an experimental and a control group was chosen. Based on the first two stages, the statistical model has to be selected. A t-test turned out to be adequate. Such statistical tests should fulfill some criteria. Usually, the test statistic should be estimated as precisely as possible. For the t-test the variance of the difference between the means of the experimental and control group should be as small as

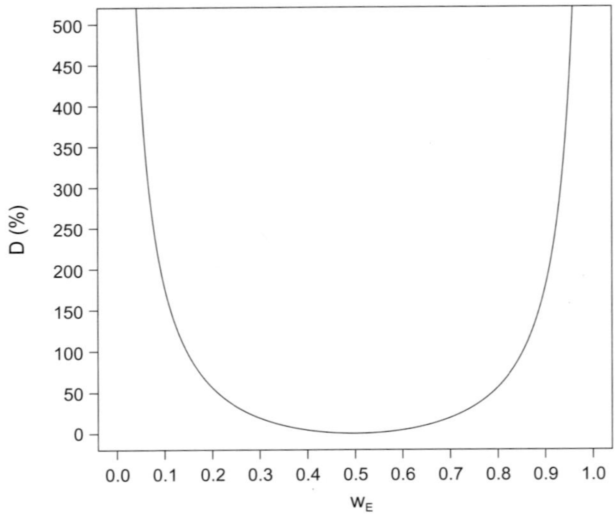

Figure 5. Graph of the deficiency for approximate designs of the prestudy.

possible. This variance is influenced by the sample size, but, independent of the sample size, also by the proportion of the subjects assigned to the experimental (and thus the control) group or, in other words, it is affected by the weights for the values of the independent variable. Thus, for any sample size the assignment of the subjects to the values of the independent variable should be optimized. If, furthermore, the sample size is not fixed because of, for example the budget, it may be determined such that a certain level of the variance of the statistic to be estimated will be achieved. The smaller the variance of the statistic, the higher is the power of a test.

We will see below that instead of the variance of the test statistic, more complex criteria are to be optimized for models with more than one parameter. It should be mentioned that the first three steps just listed above strongly determine the quality of the results of a study. We have to consider that no data or any analyses of data are involved so far.

Optimal Design for Simple Regression

In this section optimal designs for the simple linear regression will be outlined. Again, we start with an example from dyscalculia research.

Children with dyscalculia as well as children with MLD often show a distorted or imprecise representation of the number line. Such a representation can be investigated by the computer-based number line paradigm. Here, the child's task is to locate numbers (e.g., ranging from 2 to 98), shown above the number line by clicking on a specific position using the mouse (see Figure 6). Here, for the sake of simplicity we assume that the independent variable can range from 0 to 100.

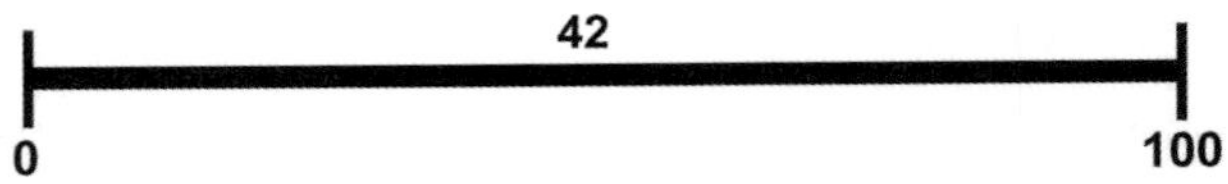

Figure 6. Number line task.

Both children with dyscalculia and MLD tend to overrepresent the spatial distance of small numbers. For children with dyscalculia a nonlinear function will result, while for children with MLD and for typically achieving children a linear relationship may be assumed (see, e.g., Geary, Hoard, Nugent, & Byrd-Craven, 2008; Kuhn et al., 2013).

Again the question arises: Which is the optimal design, that is which numbers to be presented should be selected? In addition, the distribution of the values of the independent variables has to be specified.

In number line experiments, often nearly equidistantly distributed numbers are presented. We also used such a distribution for a total of 23 numbers presented to each member of three groups: children with dyscalculia, children with MLD, and typically achieving children (control group).

For the group of children with MLD and for the control group, simple linear regression has been applied to analyze the relationship between the presented numbers (independent variable, *X*) and the estimated numbers (dependent variable, *Y*). According to the usual approach in dyscalculia research, we use the groupwise median of the estimated number for each value of the presented number as dependent variable.

In Figure 7 the results for the control group are depicted, furthermore Figure 7 contains the results for children with MLD.

Two important differences can be observed between the results of both groups: First, children with MLD underestimate the presented number by a certain amount depending on the level of the presented number. Thus, the slope of the regression line is lower than that of control children. Second, the regression line of children without dyscalculia nearly goes through the values 0 of the dependent and independent variable.

Therefore, as a statistical model for children with MLD, simple linear regression, including an intercept, is adequate:

$$Y_i = \beta_0 + \beta_1 x_i + \varepsilon_i \quad i = 1, \ldots, n, \tag{11}$$

or

$$\mathrm{E}(Y_i) = \beta_0 + \beta_1 x_i. \tag{12}$$

Here, Y_i represents the response for the *i*th presented number and x_i the known value of the *i*th predictor, that is the presented number. Moreover, as usual we assume the errors ε_i to be independently identically distributed according to normal distribution with constant variance σ^2, that is $\varepsilon_i \sim N(0, \sigma^2)$.

For typically achieving children we assume the analogous model without an intercept:

$$E(Y_i) = \beta_1 x_i. \tag{13}$$

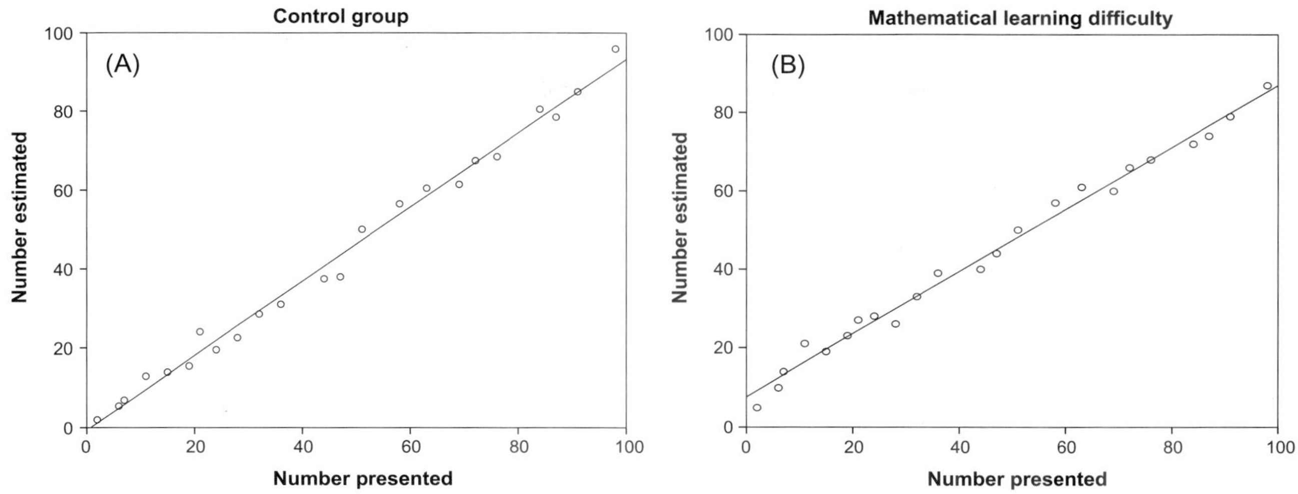

Figure 7. Scattergram for the relationship between the presented and estimated numbers for the control group (A) and children with MLD (B).

The numeric results for both groups are presented in Figures 8 and 9. It should be mentioned that in simple regression with an intercept for typically achieving children, the estimated intercept is, as expected, not significant (p = .53).

The simple linear regression model without an intercept contains one parameter β_1. As for the t-test we would like to minimize the variance of

$$\operatorname{Var}\left(\hat{\beta}_1\right) = \frac{\sigma^2}{\sum x_i^2}. \tag{14}$$

The variance of the errors is usually unknown and will be estimated by

$$\hat{\sigma}_\epsilon^{\,2} = \frac{\Sigma(y_i - \hat{\beta}_1 x_i)^2}{n-1}. \tag{15}$$

Obviously, $\operatorname{Var}(\hat{\beta}_1)$ is minimized if Σx_i^2 is maximized. Thus, only the maximum value of the predictor should be used to obtain a small variance of $\hat{\beta}_1$.

When $\operatorname{Var}(\hat{\beta}_1)$ is minimized R^2, which measures the amount of explained variation, is maximized. Furthermore, the test statistic $t = \hat{\beta}_1/SE(\hat{\beta}_1)$ is one-to-one related to the F-statistic, which determines the p-value of R^2, by $F = t^2$. In our example with the given (equidistant) design F = 2,245.754 and t = 47.495 (see Figure 8).

The simple linear regression model including an intercept contains two parameters β_0 and β_1. Therefore, several criteria for a precise estimation of parameters can be specified, for example minimizing the variance of $\hat{\beta}_0$, minimizing the variance of $\hat{\beta}_1$, or minimizing the mean variance of both parameters. We will later return to the problem of multiple optimality criteria. Mostly, a precise estimation of β_1 is of interest in simple linear regression. Again, we

Model Summary

Model	R	R Square	Adjusted R Square	Std. Error of the Estimate
1	.995	.990	.990	4.98078

ANOVA

Model		Sum of Squares	df	Mean Square	F	Sig.
1	Regression	55961.219	1	55961.219	2255.754	.000
	Residual	545.781	22	24.808		
	Total	56507.000	23			

Coefficients

Model		Unstandardized Coefficients		Standardized Coefficients	t	Sig.
		B	Std. Error	Beta		
1	Number presented	.912	.019	.995	47.495	.000

Figure 8. Results for the linear regression with an intercept of the estimated on the presented numbers for typically achieving children (control group).

Model Summary

Model	R	R Square	Adjusted R Square	Std. Error of the Estimate
1	.995	.989	.989	2.52696

ANOVA[a]

Model		Sum of Squares	df	Mean Square	F	Sig.
1	Regression	12633.382	1	12633.382	1978.440	.000
	Residual	134.096	21	6.386		
	Total	12767.478	22			

Coefficients[a]

Model		Unstandardized Coefficients		Standardized Coefficients	t	Sig.
		B	Std. Error	Beta		
1	(Constant)	7.735	.963		8.029	.000
	Number presented	.793	.018	.995	44.480	.000

Figure 9. Results for the linear regression with an intercept of the estimated on the presented numbers for children suffering from MLD.

consider a minimal variance of $\hat{\beta}_1$ as criterion. The variance of $\hat{\beta}_1$ in this regression model is

$$\operatorname{Var}\left(\hat{\beta}_1\right) = \frac{\sigma^2}{\sum (x_i - \bar{x})^2}. \tag{16}$$

Thus, the denominator $\sum (x_i - \bar{x})^2$ has to be maximized. The maximum is achieved when half of the n values of the independent variables are equal to 0 and the other half to 100. For an odd sample size $n + 1$, we have to take $(n - 1)/2$ and $(n + 1)/2$ for either of the extreme predictor values. Therefore, only the most extreme values of the independent variable have to be selected to get an optimal design.

In simple regression with an intercept, R^2 is directly related to $\hat{\beta}_1$ only, but not to $\hat{\beta}_0$, and F corresponds only to the t-statistic of $\hat{\beta}_1$.

Table 1 contains the efficiency and deficiency for other designs with equidistantly distributed values of the independent variable and identical weights in each case.

For instance, if we took three equidistantly distributed values of the independent variable with the same weight 1/3, which corresponds to the selection of the values 0, 50, and 100 each with the weight 1/3, the deficiency would be 50%. Therefore, half of the sample size of the optimal design has to be additionally investigated in order to get the same variance of the slope estimator as when using the optimal design.

According to the common practice in the number line experiment, 23 nearly equidistant design points were chosen. Compared to an optimal design we obtain an efficiency of 36%, that is 175% observations have to be additionally included to get the same value of $\operatorname{Var}(\hat{\beta}_1)$, while applying the optimal design.

If one observes, for example, five equidistantly distributed values of the independent variable with the same weight 1/5, which corresponds to the selection of the values 0, 25, 50, 75, and 100 each with the weight 1/5, the efficiency would be 50% of an optimal design. Then, the sample size has to be twice as large as that of the optimal design. Compared to the optimal design, 100% more observations have to be investigated in order to get the same variance of the slope estimator.

Table 1. Efficiency (E) and deficiency (D) for simple regression designs defined by the number of equidistantly distributed values of the independent variable (M) and identical weights in each case

M	E	D(%)
3	.67	50
4	.56	80
5	.50	100
7	.44	125
10	.41	145
23	.36	175
100	.34	194
10,000	.33	200

In the number line experiment it does not seem to be reasonable to present the same numbers repeatedly. Thus, instead of applying the optimal design consisting of the most extreme values of the predictor, one would present numbers near 10 and 90 with a loss of efficiency that would not be that high. However, in many experimental settings, only the experimental units with the most extreme values of the independent variable can be selected. For example, in the mental rotation paradigm (Shepard & Metzler, 1971) where the optimal design specifies the degrees of rotation of the stimuli to be selected, only the stimuli with the most extreme degrees could be presented for several times.

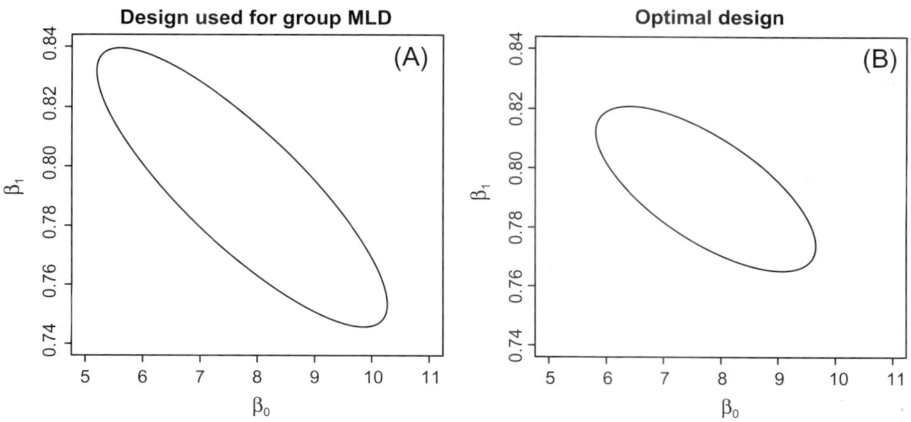

Figure 10. 95%-confidence ellipse for β_0 and β_1 of the simple regression model based on the design and data for the group with MLD (A) and based on the optimal design and parameter estimates for the group with MLD (B).

In the present example, the range of independent variables varies from 0 to 100. In general, the range of independent variables in simple regression as well as other statistical methods differs from study to study. It is favorable for specifying some optimal designs to have a unique range for the independent variables. Thus, the range of independent variables is often rescaled to an interval from -1 to $+1$ according to the transformation

$$x' = 2\frac{x - x_{\min}}{x_{\max} - x_{\min}} - 1. \tag{17}$$

Thus, the optimal design for a minimal $\mathrm{Var}(\hat{\beta}_1)$ in simple regression with an intercept is often specified for values of the predictor ranging from -1 to 1 since the points corresponding to the optimal design can be easily transformed to any other interval of design points. Such a transformation is also possible for many other optimal designs but not for the optimal design based on minimizing $\mathrm{Var}(\hat{\beta}_0)$ in simple regression. The variance of $\hat{\beta}_0$ is computed by

$$\mathrm{Var}\left(\hat{\beta}_0\right) = \sigma^2 \frac{\sum x_i^2}{n\sum (x_i - \bar{x})^2}. \tag{18}$$

The optimal design based on a minimal $\mathrm{Var}(\hat{\beta}_1)$ is neither affected by the scaling of x, that is by affine transformations $x' = bx + a$ of the predictor nor by reparameterization of β_0 and β_1, such as $\beta'_0 = 2\beta_0$. In any case, the optimal design consists of the two extreme values of the predictor combined with a weight of .5 for either value.

The optimal design based on a minimal $\mathrm{Var}(\hat{\beta}_0)$ is, however, affected by scaling and reparameterization. Using a scale from 0 to 100 results in a different optimal design for a minimal $\mathrm{Var}(\hat{\beta}_0)$, here corresponding to σ^2/n, than using predictor values ranging from -1 to 1. Therefore, attention has to be paid to the scale used in a study.

Furthermore, the optimal design for a minimal $\mathrm{Var}(\hat{\beta}_0)$ is not unique. Given a scale from –1 to +1, $\mathrm{Var}(\hat{\beta}_0)$ is minimized by any design with $\bar{x} = 0$. The exact optimal design based on a minimal $\mathrm{Var}(\hat{\beta}_0)$ for our number line example – given $n = 23$ and a scale from 0 to 100 – consists of the value 0 with a weight of 1. The drawback of this design is that β_1 cannot be estimated any longer.

Instead of minimizing either $\mathrm{Var}(\hat{\beta}_0)$ or $\mathrm{Var}(\hat{\beta}_1)$, further criteria can be considered. One might minimize the mean of the variances of $\hat{\beta}_0$ and $\hat{\beta}_1$. This criterion which is equivalent with minimizing the sum of variances of $\hat{\beta}_0$ and $\hat{\beta}_1$ is called A-optimality.

The criterion used most often in applications is D-optimality, which will be outlined now. Minimizing $\mathrm{Var}(\hat{\beta}_0)$ and $\mathrm{Var}(\hat{\beta}_1)$ corresponds to minimizing the lengths of the confidence intervals of β_0 and β_1. But, here the confidence intervals are considered separately, leading to other conclusions than a confidence ellipse for both parameters together which is based on the whole covariance matrix of $\hat{\beta}_0$ and $\hat{\beta}_1$

$$\begin{pmatrix} \mathrm{Var}\left(\hat{\beta}_0\right) & \mathrm{Cov}\left(\hat{\beta}_0, \hat{\beta}_1\right) \\ \mathrm{Cov}\left(\hat{\beta}_0, \hat{\beta}_1\right) & \mathrm{Var}\left(\hat{\beta}_1\right) \end{pmatrix}. \tag{19}$$

In simple regression this covariance matrix turns out to be

$$\sigma^2 \begin{pmatrix} \frac{\sum x_i^2}{n\sum (x_i - \bar{x})^2} & -\frac{\sum x_i}{n\sum (x_i - \bar{x})^2} \\ -\frac{\sum x_i}{n\sum (x_i - \bar{x})^2} & \frac{1}{\sum (x_i - \bar{x})^2} \end{pmatrix}. \tag{20}$$

The estimators $\hat{\beta}_0$ and $\hat{\beta}_1$ can be written as a vector

$$\hat{\boldsymbol{\beta}} = \begin{pmatrix} \hat{\beta}_0 \\ \hat{\beta}_1 \end{pmatrix} \tag{21}$$

and the covariance matrix as of $\mathrm{Cov}(\hat{\boldsymbol{\beta}})$.

Figure 10 displays the 95%- confidence ellipse for β_0 and β_1 of the simple regression model based on the data for the group with MLD in the number line experiment. This confidence ellipse contains both parameters simultaneously and takes the variances $\hat{\beta}_0$ and $\hat{\beta}_1$ into account as well as the covariance of $\hat{\beta}_0$ and $\hat{\beta}_1$.

The area of this confidence ellipse is measured by the determinant of this covariance matrix. (We will not go into details since the computation of the determinant is a part of matrix algebra.) A D-optimal design is given if the determinant of $\mathrm{Cov}(\hat{\boldsymbol{\beta}})$ is minimal where $\hat{\boldsymbol{\beta}}$ is the vector consisting of $\hat{\beta}_0$ and $\hat{\beta}_1$. As outlined when minimizing $\mathrm{Var}(\hat{\delta})$ (see above), it is favorable to consider a convex function which is equivalent to minimizing log $\det(\mathrm{Cov}(\hat{\boldsymbol{\beta}}).)$ This function is minimized to determine a D-optimal design. Figure 10 depicts the 95%-confidence ellipse for β_0 and β_1 of the simple regression model based on the optimal design and the estimated parameters for the group with MLD. The area of this confidence ellipse is much smaller than that one based on the design used in the experiment.

The criteria described above rest upon the covariance matrix of $\hat{\beta}_0$ and $\hat{\beta}_1$ or at least on a part of this matrix. Two further popular optimality criteria are based on the accuracy of the predicted responses. In simple regression, the response should be well predicted. For a certain value x_a of the predictor the predicted response of Y_a is given by $\hat{\beta}_0 + \hat{\beta}_1 x_a$. Confidence or so-called prediction intervals for these predicted responses are given by

$$\left(\hat{\beta}_0 + \hat{\beta}_1 x_a\right) \mp z_{1-\frac{\alpha}{2}}\, \sigma \sqrt{1 + \frac{1}{n} + \frac{(x-\bar{x})^2}{\sum (x_i - \bar{x})^2}}. \quad (22)$$

The prediction intervals should be as close as possible to the regression line and depend on

$$\mathrm{Var}\left(\hat{Y}_a - \left(\hat{\beta}_0 + \hat{\beta}_1 x_a\right)\right) = \sigma^2 \left(1 + \frac{1}{n} + \frac{(x-\bar{x})^2}{\sum (x_i - \bar{x})^2}\right). \quad (23)$$

Figure 11 displays the 95%-prediction intervals for the estimated numbers for children with MLD in the number line experiment.

Closely related to the prediction interval of single values of the predictors is the prediction interval of $\widehat{E(Y)}$, the expectation of the response Y. $\widehat{E(Y)}$ corresponds to the values of the regression line. It is usually called predicted response and denoted by $\hat{Y}(x)$. The variance of the predicted responses or, in other words, the variance of the regression line is given by

$$\mathrm{Var}\left(\hat{Y}(x)\right) = \sigma^2 \left(\frac{1}{n} + \frac{(x-\bar{x})^2}{\sum (x_i - \bar{x})^2}\right). \quad (24)$$

The smaller this variance, the closer are the prediction intervals for $\hat{Y}(x)$ to the regression line. Figure 11 shows the 95%-prediction intervals for the regression line for the children with MLD.

Since $\mathrm{Var}(\hat{Y}(x))$ depends on the sample size and the unknown error variance σ^2, the standardized variance will be considered for developing optimality criteria. The standardized variance is adjusted for the sample size as well as for the error variance σ^2 and defined for exact designs by

$$\begin{aligned} \mathrm{d}(x, \xi_n) &= \frac{n}{\sigma^2}\mathrm{Var}\left(\hat{Y}(x)\right) \\ &= \frac{n}{\sigma^2}\sigma^2\left(\frac{1}{n} + \frac{(x-\bar{x})^2}{\sum (x_i - \bar{x})^2}\right) \\ &= 1 + n\frac{(x-\bar{x})^2}{\sum (x_i - \bar{x})^2}. \end{aligned} \quad (25)$$

For approximate designs the standardized variance is defined by

$$\mathrm{d}(x, \xi) = 1 + \frac{(x-\bar{x})^2}{\sum w_i (x_i - \bar{x})^2}. \quad (26)$$

Thus, the standardized variance does not incorporate the error variance, which is usually unknown, and it is independent of the number of the experimental runs.

Figure 12 shows the standardized variance for the experimental design used in the number line experiment. Here, 23 nearly equidistantly distributed values of the predictor were presented. The interval between 0 and 20 contains more design points than the interval between 80 and 100. Therefore, the standardized variance is smaller for design points near $x = 10$ than $x = 90$.

The D-optimal design consists of the values 0 and 100, each with a weight of .5. The standardized variance for this design, which is mostly smaller over the design region, is displayed in Figure 12. Thus, the responses are better predicted when using the D-optimal design.

The standardized variance is a function defined on the whole design space, but not a single value. In order to compare different standardized variances, it is, in general, necessary to summarize the values of this function. Two methods are usually considered, computing the maximum

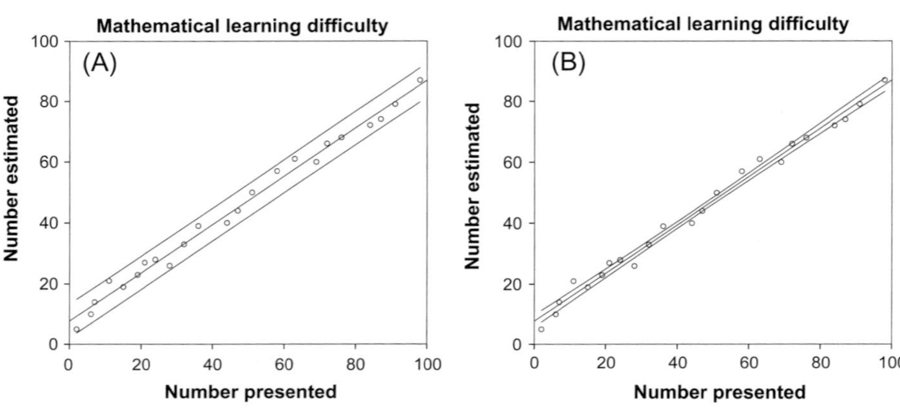

Figure 11. 95%-prediction intervals for single expected response (A) and for the regression line (B) for the estimated numbers for children with MLD in the number line experiment.

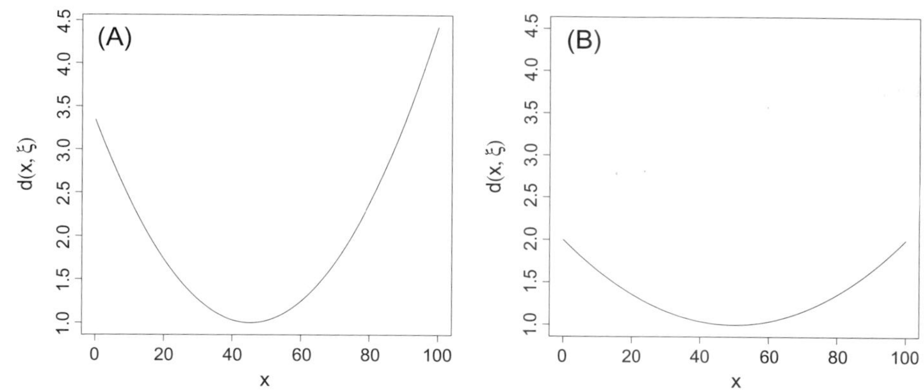

Figure 12. Standardized variance for the experimental design used in the number line experiment (A) and the corresponding optimal design (B).

or the mean of the standardized variance over the design space.

In G-optimality the maximum standardized, that is for the worst value x of the predictor under that design variance will be computed for all possible designs. Then, that design is selected yielding the smallest maximum standardized variance. Thus, a G-optimal design is given when the maximum standardized variance of the predicted response over the design region X is minimized. Here, the worst case will be prevented. In integrated mean squared error (IMSE)-optimality, also known as I-, V-, or Q-optimality, instead of the maximum the integrated mean squared error corresponding to the mean of the standardized variance is minimized.

The maximum standardized variance for the D-optimal design amounts to 2 and the maximum standardized variance for the design used in the experiment to 4.4 (see Figure 12). As we show in the next section, the D-optimal design for simple linear regression is also G-optimal.

Figure 12 shows that for the D-optimal design, which is also G-optimal, the maximal values of the standardized variance are attained at the design points 0 and 100. This equivalence is result of an important theorem which will be outlined later.

General Linear Model

The *t*-test as well as simple and multiple regression, analysis of variance and covariance, can be embedded in a more general setup, the general linear model. (In Appendix A the basics of the general linear model are outlined using matrix algebra. It is referred to the interested reader, but it is not necessary to read it to understand the following fundamentals of optimal design.)

The dependent variable in the general linear model has to be quantitative, the independent variables may be qualitative or quantitative. Yet, qualitative variables have to be transformed by, for example, dummy coding or cell mean coding (see Appendix B) to be included in the regression equation.

The general linear model is specified in the following

$$Y_i = \beta_0 + \beta_1 x_{i1} + \beta_2 x_{i2} + \cdots + \beta_p x_{ip} + \varepsilon_i, \quad (27)$$

or

$$\mathrm{E}(Y_i) = \beta_0 + \beta_1 x_{i1} + \beta_2 x_{i2} + \ldots + \beta_p x_{ip}. \quad (28)$$

Again, we assume the errors to be independently normally distributed with constant variance.

Now, the findings of the simple regression outlined above can be naturally generalized. The vector $\hat{\boldsymbol{\beta}}$ is now given by

$$\hat{\boldsymbol{\beta}} = \begin{pmatrix} \hat{\beta}_0 \\ \hat{\beta}_1 \\ \vdots \\ \hat{\beta}_p \end{pmatrix}. \quad (29)$$

The covariance matrix of the estimators $\hat{\boldsymbol{\beta}}$ can be expressed as

$$\mathrm{Cov}\left(\hat{\boldsymbol{\beta}}\right) = \begin{pmatrix} \mathrm{Var}\left(\hat{\beta}_0\right) & \mathrm{Cov}\left(\hat{\beta}_0, \hat{\beta}_1\right) & \cdots & \mathrm{Cov}\left(\hat{\beta}_0, \hat{\beta}_p\right) \\ \mathrm{Cov}\left(\hat{\beta}_1, \hat{\beta}_0\right) & \mathrm{Var}\left(\hat{\beta}_1\right) & \cdots & \mathrm{Cov}\left(\hat{\beta}_1, \hat{\beta}_p\right) \\ \vdots & \vdots & \vdots & \vdots \\ \mathrm{Cov}\left(\hat{\beta}_p, \hat{\beta}_0\right) & \mathrm{Cov}\left(\hat{\beta}_p, \hat{\beta}_1\right) & \cdots & \mathrm{Var}\left(\hat{\beta}_p\right) \end{pmatrix}. \quad (30)$$

The predicted response $\hat{Y}(x_1, \ldots, x_p)$ is now given as $\hat{\beta}_0 + \sum_{i=1}^{p} \hat{\beta}_i x_i$. Again, its variance will be of interest in specifying optimality criteria.

The optimality criteria, as introduced for simple regression, can be naturally generalized (for a comprehensive overview, see Atkinson, Donev, & Tobias, 2007). To keep it simple, we will omit advanced mathematical prerequisites, which are, for example, necessary for the proper definition of the criteria.

First, the variance of single estimators of parameters, $\mathrm{Var}(\hat{\beta}_i)$, $i = 0, \ldots, p$ can be minimized. Again, in

A-optimality the mean variance of all estimators $\hat{\beta}_i$, $i = 0, \ldots, p$ or equivalently $\sum_{i=0}^{p} \text{Var}(\hat{\beta}_i)$ is minimized.

The most frequently used criterion, D-optimality, is given as the minimum of log det(Cov($\hat{\boldsymbol{\beta}}$)) which corresponds to as small as possible volumes of the confidence ellipsoids containing the regression coefficients $\beta_0, \beta_1, \ldots, \beta_p$. The determinant of Cov($\hat{\boldsymbol{\beta}}$) is also referred to as the generalized variance.

As in simple regression, G-optimality and IMSE-optimality minimize the maximal variance and mean variance of the predicted response, respectively.

Hypotheses may only refer to subsets of the parameters. Concerning the line number experiment, specific hypotheses may only be related to the slope of the regression line. When the volume of the confidence ellipsoid should be minimized for a specific subset of parameters, D_s-optimality is the appropriate criterion. Here, D-optimality is limited to a subset of $s < p$ parameters and the minimum of log det($\text{Cov}(\hat{\boldsymbol{\beta}})_s$) has to be determined where $\text{Cov}(\hat{\boldsymbol{\beta}})_s$ is the submatrix of Cov($\hat{\boldsymbol{\beta}}$) containing the parameters of interest. The other parameters are then treated as nuisance parameters.

Often, special linear combinations of parameters are of interest. Thus, the training success of several groups may be compared as in our example concerning the trainings for reducing dyscalculia *the training success of the two experimental groups – the training success of the two control groups* may be analyzed.

In c-optimality the variance of the estimator of a single specific linear combination of the parameters $\sum_{i=0}^{p} c_i\hat{\beta}_i$ is minimized. Often, hypotheses refer to a certain combination of means. Such hypotheses are simply specified using a cell mean coding (see Appendix B). If a cell mean coding is applied and no explicit intercept is included in the regression equation, the regression coefficients correspond to the mean of certain groups, for example $\beta_1 = \mu_{E1}$, $\beta_2 = \mu_{E2}$, $\beta_3 = \mu_{C1}$, and $\beta_4 = \mu_{C2}$. When the mean of both experimental groups, which is $.5\ \mu_{E1} + .5\ \mu_{E2}$, should be compared to the mean of both control groups corresponding to $.5\ \mu_{C1} + .5\ \mu_{C2}$ the variance of the estimator of the linear combination $(\beta_1 + \beta_2) - (\beta_3 + \beta_4)$ has to be minimized.

More generally, when several linear combinations of the parameters will be simultaneously investigated, D_A-optimality is the appropriate criterion. For example, in addition to the above linear combination $(\beta_1 + \beta_2) - (\beta_3 + \beta_4)$, a second one concerning the difference between both experimental groups $\mu_{E1} - \mu_{E2} = \beta_1 - \beta_2$ might be of interest in order to determine which intervention is more efficient. Then, two linear combinations $(\beta_1 + \beta_2) - (\beta_3 + \beta_4)$ and $\beta_1 - \beta_2$ have to be considered and the corresponding generalized variance of both linear combinations is simultaneously minimized (see Appendix C).

The minimization of the variance of single estimators of parameters, Var($\hat{\beta}_i$) can be identified as c-optimality. Here, the $c_i = 1$ for the interesting parameter β_i and 0 elsewhere.

A major advantage of D-optimality and as well of the G- and IMSE-optimality compared to other optimality criteria is the independence of the optimal design from the parameterization chosen and from the scaling of the predictors. When the scale of a predictor is transformed by an affine transformation, the same D-optimal design will result. But, scale dependence might not be a disadvantage if a "natural" scale for the predictor is given, for example for our line number experiment.

The construction of optimal designs is based on the optimization of one of the criteria outlined above. Thus, optimization theory and especially convex analysis is an important tool for developing optimal designs. However, often these optimization problems cannot be solved analytically. The minimization of Var($\hat{\beta}_i$) in the model $E(Y) = \beta_0 + \beta_1 x + \beta_2 x^2$ is already a very complicated issue, as shown in Appendix D.

When the optimal design cannot be derived analytically, it has to be determined numerically. A lot of algorithms have been developed that provide optimal or nearly optimal designs. One classical algorithm, the Fedorov algorithm, is outlined in Appendix E. This approach and further algorithms are meanwhile part of standard software, for example in SAS (see Atkinson et al., 2007).

The Fedorov exchange algorithm and other related algorithms to determine optimal designs are based on the so-called equivalence theory that provides fundamental theorems in optimal design theory. Here, we shortly describe an equivalence theorem for the general linear model (see Kiefer & Wolfowitz, 1960).

According to this theorem a G-optimal design that minimizes the maximal standardized variance is also D-optimal. For polynomial regression models, the number of distinct design points corresponds to $p + 1$, the number of parameters to be estimated. The design points are given for those values that attain the maximum value for the standardized variance. Furthermore, the maximal standardized variance amounts to $p + 1$.

Figure 12(B) shows these relationships for simple regression. The D-optimal design is G-optimal and the maximal values of the standardized variance are attained at the design points 0 and 100. This equivalence can also be shown for the quadratic model $E(Y) = \beta_0 + \beta_1 x + \beta_2 x^2$. The standardized variance for a quadratic regression model given the design

$$\left\{ \begin{array}{ccc} 0 & 50 & 100 \\ 1/3 & 1/3 & 1/3 \end{array} \right\} \tag{31}$$

is depicted in Figure 13. According to the equivalence theorem outlined above, the maximum of the standardized variance, which amounts to 3, is attained for the design points 0, 50, and 100. Thus, this design is D-optimal as well as G-optimal.

For quadratic regression on the standardized design region $[-1, 1]$, A-optimal designs, IMSE-optimal designs as well as the c-optimal design for minimizing the variance of $\hat{\beta}_2$ use in accordance with D- and G-optimal designs the two extreme values and the midpoint of a scale as designs points. But the weights, which are presented in Table 2, differ.

A cubic model $E(Y) = \beta_0 + \beta_1 x + \beta_2 x^2 + \beta_3 x^3$ is appropriate to model the relationship between the presented and

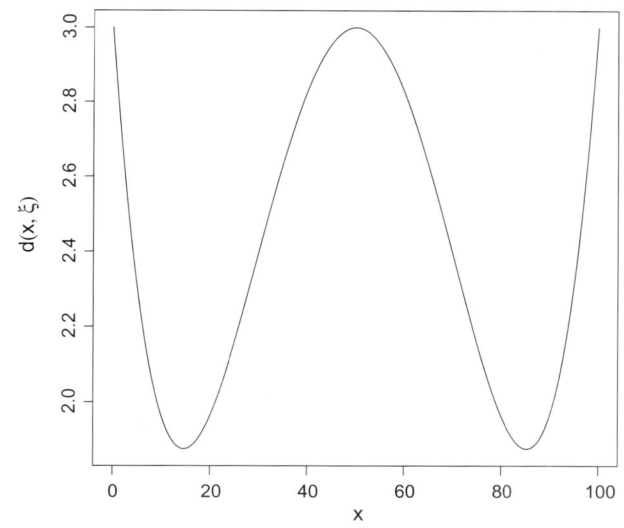

Figure 13. Standardized variance for the quadratic regression given the D-optimal design for a continuous predictor with a range from 0 to 100.

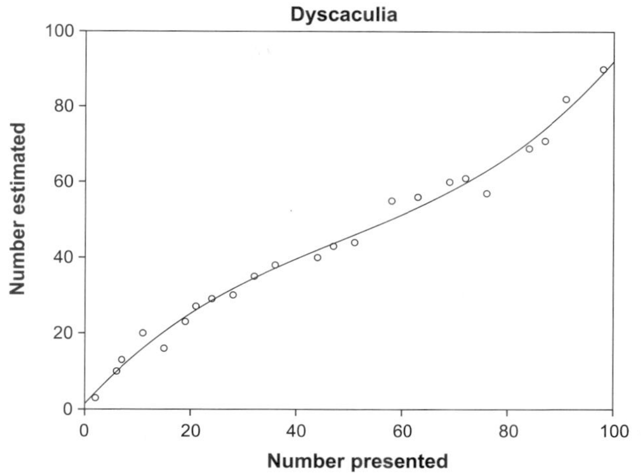

Figure 14. Scattergram and cubic regression line for the relationship of the presented and estimated numbers for children with dyscalculia.

estimated numbers in the number line experiment for children with dyscalculia (see Figure 14). Here, the D-optimal design is

$$\left\{ \begin{array}{cccc} 0 & 28 & 72 & 100 \\ 1/4 & 1/4 & 1/4 & 1/4 \end{array} \right\}. \tag{32}$$

Figure 15 displays the standardized variance for the cubic regression model given the D-optimal design.

Table 3 contains the D-optimal designs for all polynomial regression models $E(Y) = \beta_0 + \sum_{i=1}^{s} \beta_i x_i^s$ up to the degree $s = 5$ on the standardized region $\{-1, 1\}$. The optimal weights are $1/(s + 1)$ in any case. Considering for instance cubic regressions, this means, the optimal design points consist of -1, -0.447, .447, and 1, each with a weight of .25.

Optimal design theory for linear models is well developed. The results outlined for the *t*-test can be generalized to analyzing mean differences of more than two groups by a one-way ANOVA. Here, D-optimal designs are given if all groups consist of the same number of experimental units. In simple regression D-optimal designs were given by the minimal and maximal values of the predictor, each with a weight of .5. For general linear models including an intercept and with more than one predictor D-optimal design consists of all possible combinations of the extreme values of the predictors with equal weights. When, for instance, two predictors x_1 and x_2 each with minimal value -1 and maximal value 1 are included in the regression equation, the D-optimal design is given by the pairs (x_1, x_2) of values of x_1 and x_2: $(-1, -1)$, $(-1, +1)$, $(+1, -1)$, and $(+1, +1)$, each with a weight of .25. Therefore, full factorial designs are D-optimal for ANOVA and multiple regression.

There is a comprehensive literature which provides optimal designs for many specific optimality criteria and linear models, as, for example, with homoscedastic, heteroscedastic, or correlated errors (see Atkinson et al., 2007; Berger & Wong, 2009; Fedorov & Hackl, 1997).

The development of optimal design for nonlinear models is much more complicated as we will show in the next section.

Optimal Design for Nonlinear Models

In the general linear model the covariance matrix $\mathrm{Cov}(\hat{\boldsymbol{\beta}})$ does not contain the parameters to be estimated. Thus, the optimal design can be derived independently of the values of the true parameters. However, in nonlinear models this covariance matrix in general encloses the true parameters whose values are to be estimated. Therefore, specific values of the parameters have to be assumed for determining an optimal design. Optimal designs based on specific values of the parameters are called locally optimal designs. When good guesses of the values of the parameters are available, locally optimal designs may be used, especially if the optimal designs are robust, that is, they do not depend too much on the parameter values. Furthermore, different

Table 2. Weights of the D-, G-, A-, and IMSE-optimal designs as well as c-optimal design for minimizing the variance of $\hat{\beta}_2$

	Design points		
Criterion	-1	0	1
D, G	.333	.333	.333
A, IMSE, min $\mathrm{Var}(\hat{\beta}_2)$	.25	.50	.25

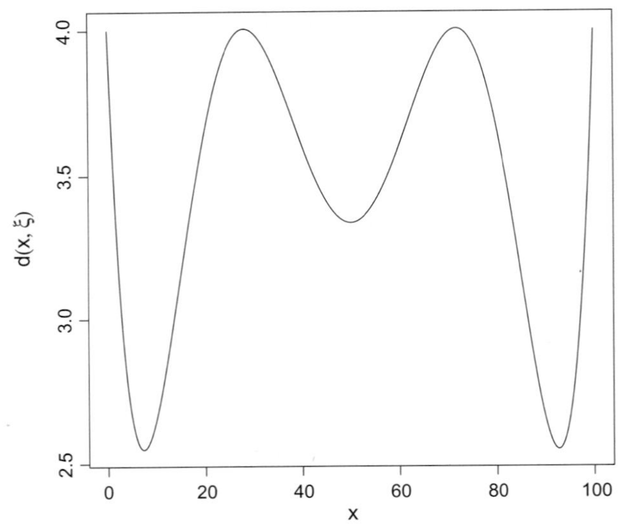

Figure 15. Standardized variance for the cubic regression given the D-optimal design for a quantitative predictor with a range from 0 to 100 according the number line experiment.

strategies can be applied to overcome the dependence of the parameters. The methods mostly used are the development of maximin, adaptive, and Bayesian designs. These strategies all rely on locally optimal designs which in any case have to be derived first.

In this introductory article we will only give a very short insight into the development of locally optimal designs for nonlinear models. As an illustrative example we will use the evaluation of an intervention for reducing dyscalculia, which is measured by a binary diagnosis. Some standard procedures for measuring dyscalculia result in such diagnoses. Using a binary response as the dependent variable, logistic regression is an adequate statistical model for comparing the outcomes of an intervention group and a control group. The independent variable is here binary, too. The design problem now concerns the proportion of respondents assigned to the experimental (intervention) group.

Let Y be a binary variable with $Y = 1$ if the event is given and $Y = 0$ if the event is not given. The logistic regression is specified by

$$P(Y = 1) = \frac{\exp(\beta_0 + \beta_1 x)}{1 + \exp(\beta_0 + \beta_1 x)}. \tag{33}$$

For the provided example, let $Y = 1$ if dyscalculia is given and $Y = 0$ if dyscalculia is not given and, furthermore, $x = 0$ for members of the control group and $x = 1$ for members of the experimental group. Thus, the following regression equations result

$$P(Y = 1;\, x = 0) = p_0 = \frac{\exp(\beta_0)}{1 + \exp(\beta_0)}, \tag{34}$$

$$P(Y = 0; x = 0) = 1 - p_0 = \frac{1}{1 + \exp(\beta_0)}, \tag{35}$$

$$P(Y = 1; x = 1) = p_1 = \frac{\exp(\beta_0 + \beta_1)}{1 + \exp(\beta_0 + \beta_1)}, \tag{36}$$

$$P(Y = 0; x = 1) = 1 - p_1 = \frac{1}{1 + \exp(\beta_0 + \beta_1)}. \tag{37}$$

The odds for the ratio of the probability to suffer from dyscalculia versus not to suffer from dyscalculia is given for the experimental group by

$$\text{odds}\,(Y = 1;\, x = 1) = \frac{p_1}{1 - p_1} = \exp(\beta_0 + \beta_1) \tag{38}$$

and for the control group by

$$\text{odds}\,(Y = 1;\, x = 0) = \frac{p_0}{1 - p_0} = \exp(\beta_0). \tag{39}$$

The odds ratio (OR)

$$\text{OR} = \frac{\text{odds}\,(Y = 1;\, x = 1)}{\text{odds}\,(Y = 1;\, x = 0)} \tag{40}$$

is given by

$$\text{OR} = \exp(\beta_1). \tag{41}$$

The lower the odds ratio, the higher is the success of the intervention. Therefore, the regression coefficient β_1 is of interest and $\text{Var}(\hat{\beta}_1)$ should be minimized in dependence of the proportion of respondents in the experimental (or control) group.

It can be shown (see e.g., Dette, 2004) that the asymptotic variance is given by

Table 3. Design points of D-optimal designs for polynomial regression models up to the degree of 5 in the design region [−1, 1]

Degree	Design points					
1	−1	1				
2	−1	0	1			
3	−1	−0.447	0.447	1		
4	−1	−0.665	0	0.665	1	
5	−1	−0.765	−0.285	0.285	0.765	1

$$\operatorname{Var}\left(\hat{\beta}_1\right) = \frac{1}{n_E p_1(1-p_1)} + \frac{1}{n_C p_0(1-p_0)}, \tag{42}$$

with n_E and n_C as sample sizes of the experimental and control group, respectively, and $n_E + n_C = n$.

Using the regression Equations 34–36 $\operatorname{Var}(\hat{\beta}_1)$ can be expressed as

$$\operatorname{Var}\left(\hat{\beta}_1\right) = \frac{1}{n_E \frac{\exp(\beta_0+\beta_1)}{(1+\exp(\beta_0+\beta_1))^2}} + \frac{1}{n_C \frac{\exp(\beta_0)}{(1+\exp(\beta_0))^2}}. \tag{43}$$

Here, the variance of $\hat{\beta}_1$ depends upon the true parameters β_0 and β_1 or, equivalently, on the conditional probabilities p_0 and p_1. Thus, in order to determine the optimal design, specific values of the parameters, which might come from previous studies, have to be assumed.

The locally optimal design based upon specific parameter values is obtained by minimizing $\operatorname{Var}(\hat{\beta}_1)$. Dette (2004) proved that the minimal $\operatorname{Var}(\hat{\beta}_1)$ is given by the following weights w_1 for $x = 1$ and w_0 for $x = 0$:

$$w_1 = \frac{\sqrt{p_0(1-p_0)}}{\sqrt{p_0(1-p_0)} + \sqrt{p_1(1-p_1)}}, \tag{44}$$

and

$$w_0 = 1 - w_1. \tag{45}$$

Table 4 shows the weights of the optimal design in dependence of some true parameters.

A first way to cope with the problem that the design depends on the parameters is the development of adaptive or sequential designs. There are many different kinds of adaptive designs. Their common core can be described as follows.

Adaptive designs consist of several stages. In the first stage a locally optimal design, for example based on a guess of the parameters, will be used to preliminarily estimate the parameters. In the second stage a new locally optimal design utilizing the parameter estimates from the first stage as values for the local optimality is applied for estimating the parameters anew. This procedure may be continued in further stages and stop after a predefined number of stages is reached or another condition such as a specific precision of the parameter estimates is fulfilled.

Adaptive designs may vary with regard to the number of stages and the experimental runs per stage. For the example presented above an adaptive design may consist of several stages, each with a certain sample size. When, for instance, the budget allows for a total sample size of $n = 300$, three stages, each with $n = 100$, may be conducted. For the first stage plausible values of the parameters p_0 and p_1 or equivalently β_0 and β_1 have to be determined, for example from previous research which allows for developing a first locally optimal design. The parameter estimates from the first stage will be used for deriving a revised design, and so on.

A reduction by using an adaptive design is shown by a small simulation study for sample sizes $n = 100$ and $n = 300$. We outline the design of the simulation study for $n = 100$, for $n = 300$ it is analogous. 100,000 samples with a sample size $n = 100$ were drawn from a population with parameters $p_0 = .8$ and $p_1 = .6$ which correspond to $\beta_0 = 1.386$ and $\beta_1 = -0.981$ as well as $w_0 = .550$ and $w_1 = .450$. The parameters were estimated by the R-routine glm for logistic regressions. For each of 10,000 samples the following three different designs were used:

1) Usual, nonadaptive design: A sample with $n = 100$ and $w_1 = w_2 = .5$ was drawn, then the parameters were estimated.
2) Adaptive design with two steps: First, a sample with $n = 50$ and $w_1 = w_2 = .5$ was drawn, then a sample with $n = 50$ using the weights based on the parameter estimates of the first sample. Then, the final parameters were estimates for the combined sample with $n = 100$.
3) Adaptive design with four steps: This design corresponds to the second one, but starts with $n = 25$. This design corresponds to the second one, but here four samples with $n = 25$ were sampled and the weights were updated after each step.

For each design the variance of 10,000 final estimates of the parameter was computed as well as the mean squared error $MSE = \sum_{i=1}^{10,000} \left(\hat{\beta}_{1i} - \beta_1\right)^2 / 10,000$. Table 5 contains $\operatorname{Var}(\hat{\beta}_1)$ and MSE observed for both sample sizes.

The results show a slight improvement for using an adaptive design. But, for other population parameters p_0 and p_1 higher gains might occur. Furthermore, there are

Table 4. Weights of the optimal designs minimizing $\operatorname{Var}(\hat{\beta}_1)$ in binary logistic regression

p_0	p_1	β_0	β_1	w_0	w_1
0.9	0.7	2.197	−1.350	0.604	0.396
0.9	0.5	2.197	−2.197	0.625	0.375
0.9	0.3	2.197	−3.045	0.604	0.396
0.7	0.5	0.847	−0.847	0.522	0.478
0.7	0.3	0.847	−1.695	0.500	0.500
0.7	0.1	0.847	−3.045	0.396	0.604
0.5	0.5	0.000	0.000	0.500	0.500
0.5	0.3	0.000	−0.847	0.478	0.522
0.5	0.1	0.000	−2.197	0.375	0.625

Table 5. $\mathrm{Var}(\hat{\beta}_1)$ and *MSE* for three designs used in the simulation study with $n = 100$ and $n = 300$ based on $p_0 = .8$ and $p_1 = .6$

	Sample size			
	100		300	
Design	$\mathrm{Var}(\hat{\beta}_1)$	*MSE*	$\mathrm{Var}(\hat{\beta}_1)$	*MSE*
Nonadaptive design	0.231	0.231	0.072	0.072
Adaptive design with two steps	0.220	0.221	0.070	0.070
Adaptive design with four steps	0.210	0.216	0.069	0.069

more sophisticated adaptive strategies yielding a higher improvement.

In Bayesian optimal designs the uncertainty about the parameters is captured by a probability distribution for the unknown parameters. Instead of assuming one parameter value, a probability distribution or prior is specified. Thus, the unknown parameters p_0 and p_1 may be described by beta distributions. Then, the optimality criterion, for example variance of the parameter estimates or the log-determinant of the information matrix, is weighted by this distribution.

This procedure will be illustrated by using a very simple discrete distribution for the parameters p_0 and p_1. We assume the following probabilities for four different pairs of values of p_0 and p_1 presented in Table 6. That means, it is believed, only four pairs of parameter values are given, each with a certain probability.

For each pair (p_0, p_1) based on the corresponding weights $\mathrm{Var}(\hat{\beta}_1)$ is now computed for a sample size of $n = 300$ as used in the simulation study. Then, the variances are weighted by the corresponding probability. The sum of these weighted variances (see the final row in Table 6) is a Bayesian criterion which has to be minimized through finding appropriate weights for the two values of the independent variable (group membership).

Usually, continuous distributions are used as priors, then integrals instead of sums have to be used in order to compute the Bayesian optimality criteria. Furthermore, for a specific optimality criterion, several different Bayesian criteria have been defined (see Atkinson et al., 2007).

A drawback of Bayesian designs is the specification of priors. Maximin designs, a further approach to overcome the dependence of optimal designs on the unknown parameters, do not require such an assumption. They only have to presuppose a range of plausible parameter values.

First, the relative efficiency for each design on the design region is determined for all parameter values over the predefined subset of the parameter space. The quality of a design is now evaluated by the minimal efficiency over all (unknown) parameter values in the set of reasonable parameter values. Finally, the design with the maximal value of the minimum efficiency is chosen. This design is called maximin efficient. Therefore, the worst case for each design is considered and the best out of the worst cases is selected.

Table 6. Probabilities for four different pairs of values of p_1 and p_0 in binary regression

p_0	p_1	$P(p_0, p_1)$	$\mathrm{Var}(\hat{\beta}_1)$	$P(p_0, p_1) \times \mathrm{Var}(\hat{\beta}_1)$
.8	.4	.20	0.069	0.014
.9	.4	.30	0.102	0.031
.8	.5	.25	0.068	0.017
.9	.5	.25	0.101	0.025
Sum				0.087

We will illustrate the development of a maximin design by using a simple example. As a subset that contains the plausible parameter values we select two discrete values .8, .9 for p_0 and three discrete values .3, .4, .5 for p_1. Furthermore, we only consider five designs defined by the weights (.3, .7), (.4, .6), (.5, .5), (.6, .4), and (.7, .3) for the two values 0 and 1 of the independent variable.

The efficiency for minimizing $\mathrm{Var}(\hat{\beta}_1)$ was defined as $\mathrm{Var}(\hat{\beta}_1^*)/\mathrm{Var}(\hat{\beta}_1)$. To compute the efficiencies for all designs given the selected pair of parameter values, first the locally optimal designs have to be determined for each pair of parameter values, then the corresponding $\mathrm{Var}(\hat{\beta}_1)$ for the optimal design denoted by $\mathrm{Var}(\hat{\beta}_1^*)$, again based on $n = 300$, can be calculated (see Table 7).

After computing $\mathrm{Var}(\hat{\beta}_1)$ for all other designs, the efficiencies $\mathrm{Var}(\hat{\beta}_1^*)/\mathrm{Var}(\hat{\beta}_1)$ are determined (see Table 8). Then, for each column, the minimum efficiency is determined and displayed in the last row of Table 8. The maximum of the values in this row yields the maximin efficient design. This maximum is 0.982, thus the maximin efficient design is given by the design with the weights .4 and .6 which is highly efficient for all parameter constellations considered.

Here, for illustrative purposes, only a few designs and pairs of true parameters have been selected. Usually, rectangular subregions $[a, b] \times [c, d]$ of the parameter space $[0, 1] \times [0, 1]$ are considered. Fortunately, for this situation, maximin efficient designs can be analytically determined. Dette (2004) proved that the weights of maximin efficient designs for a rectangular subregion (SR) of the parameter space $[0, 1] \times [0, 1]$ are given by

$$w_1 = \frac{2 + \sqrt{u_{\min}} + \sqrt{u_{\max}}}{2\left(\sqrt{u_{\min}} + \sqrt{u_{\max}}\right) + 2\left(1 + \sqrt{u_{\min}}\sqrt{u_{\max}}\right)} \tag{46}$$

and

$$w_0 = 1 - w_1, \tag{47}$$

Table 7. Weights of the optimal design and corresponding $Var(\hat{\beta}_1^*)$ for different pairs of values of p_0 and p_1 in binary regression

p_0	p_1	w_0	w_1	$Var(\hat{\beta}_1^*)$
.8	.3	0.534	0.466	0.073
.9	.3	0.604	0.396	0.101
.8	.4	0.551	0.450	0.069
.9	.4	0.620	0.380	0.096
.8	.5	0.556	0.444	0.068
.9	.5	0.625	0.375	0.095

Table 8. Efficiency of the designs for different parameters

		Design				
p_0	p_1	.3 .7	.4 .6	.5 .5	.6 .4	.7 .3
.8	.3	0.884	0.982	0.995	0.930	0.793
.9	.3	0.958	1.000	0.958	0.852	0.694
.8	.4	0.904	0.900	0.900	0.914	0.770
.9	.4	0.971	0.998	0.945	0.832	0.672
.8	.5	0.910	0.992	0.988	0.908	0.763
.9	.5	0.974	0.997	0.941	0.826	0.665
Minimum		0.884	0.982	0.900	0.826	0.665

with

$$u_{\min} = \min_{p_0, p_1 \in SR} \frac{p_1(1-p_1)}{p_0(1-p_0)},$$

$$u_{\max} = \max_{p_0, p_1 \in SR} \frac{p_1(1-p_1)}{p_0(1-p_0)}.$$

Adaptive, Bayesian, and maximin designs have been developed for many nonlinear models important in psychology. In item response theory, for instance, in particular, adaptive designs have been widely used. When the ability parameters of the so-called 1-, 2-, or 3-PL model are estimated by adaptive testing, usually the first stage contains several fixed items. The next stages, then, consist of only one item to update the parameter estimates. This iterative procedure may stop after a certain number of items have been applied or the confidence interval for the ability parameter is smaller than a predefined value.

Maximin and Bayesian optimal designs for the 1PL-model were developed by Graßhoff, Holling, and Schwabe (2012). Berger, King, and Wong (2000) provided maximin designs for the 2-PL model (for an overview, see Holling & Schwabe, in press).

Summary

This article provides an elementary introduction to optimal design. An optimal design specifies the distribution of the independent variables such that an a priori defined statistical criterion is optimized. First, basic principles of optimal design are explained using the *t*-test as a simple example. As an obvious statistical criterion, here, the most precise estimation of the mean difference of the dependent variable between the experimental and control group is chosen. It turns out that the difference of the means of the dependent variable is estimated best when half of the sample size is allocated to either group.

In simple regression often a precise estimation of the slope parameter is of interest. For minimizing the variance of the estimator of this parameter, only the minimal and maximal values of the independent variable should be selected, each with 50%.

t-Test and simple regression can be embedded in the general linear model, a more comprehensive class of statistical models. For some of these models designs that are optimal according to specific optimality criteria have been introduced.

Which criterion should be chosen depends upon the hypothesis to be investigated. When focusing on one parameter or a linear combination of parameters, minimizing the variance of the corresponding estimator will usually be the optimality criterion. In A-optimality the mean of the variance of the parameter estimators is minimized. D-optimality, which corresponds to a minimal determinant of the covariance matrix of the parameter estimates, is the most popular criterion in application. According to the general equivalence theory, it coincides with G-optimality for the general linear model. G-optimal designs achieve the maximum of the minimal standardized variance over the design space. In IMSE-optimality the integrated mean squared error corresponding to the mean of the minimal standardized variance is minimized.

The information matrix of linear models does, in general, not contain the unknown parameters. Thus, optimal designs can be developed without knowing the true parameters. Designs for nonlinear models usually depend upon the parameters to be estimated. Optimal designs then have to be derived for specific parameter values and are called locally optimal designs. To overcome the dependence of the unknown parameters adaptive, Bayesian, and maximin designs may be used.

Acknowledgments

This research was partly supported by the German Research Foundation (DFG) under Grant HO 1286/6-2 and the German Federal Ministry of Education and Research (BMBF) under Grant 01GJ1006.

References

American Psychiatric Association. (2000). *Diagnostic and statistical manual of mental disorders*. Washington, DC: Author.

Atkinson, A. C., Donev, A. N., & Tobias, R. D. (2007). *Optimum experimental designs, with SAS*. Oxford, UK: Oxford University Press.

Berger, M. P. F., King, C. Y. J., & Wong, W. K. (2000). Minimax D-optimal designs for item response theory models. *Psychometrika, 65*, 377–390.

Berger, M. P. F., & Wong, W. K. (2009). *An introduction to optimal designs for social and biomedical research*. Chichester, UK: Wiley.

Dette, H. (2004). On robust and efficient designs for risk estimation in epidemiological studies. *Scandinavian Journal of Statistics, 31*, 319–331.

Fedorov, V. V., & Hackl, P. (1997). *Model-oriented design of experiments*. New York, NY: Springer.

Geary, D. C., Hoard, M. K., Nugent, L., & Byrd-Craven, J. (2008). Development of number line representations in children with mathematical learning disability. *Developmental Neuropsychology, 33*, 277–299.

Graßhoff, U., Holling, H., & Schwabe, R. (2012). Optimal design for the Rasch model. *Psychometrika, 77*, 710–723.

Haffner, J., Baro, K., Parzer, P., & Resch, F. (2005). *Heidelberger Rechentest (HRT 1–4)*. Göttingen, Germany: Hogrefe.

Holling, H., & Schwabe, R. (in press). Optimal design theory. In W. J. van der Linden & R. K. Hambleton (Eds.), *Handbook of item response theory* (Vol. 2).

Kiefer, J., & Wolfowitz, J. (1960). The equivalence of two extremum problems. *Canadian Journal of Mathematics, 12*, 363–366.

Kuhn, J.-T., Raddatz, J., Holling, H., & Dobel, Ch. (2013). Dyskalkulie vs. Rechenschwäche: Basisnumerische Verarbeitung in der Grundschule [Dyscalculia vs. mathematical learning disability: Basic numerical processing in primary school]. *Lernen und Lernstörungen, 2*, 1–19.

Landerl, K., Bevan, A., & Butterworth, B. (2004). Developmental dyscalculia and basic numerical capacities: A study of 8–9 year old students. *Cognition, 93*, 99–125.

Lenhard, A., Lenhard, W., & Klauer, K. J. (2012). *Denkspiele mit Elfe und Mathis. Förderung des logischen Denkvermögens für das Vor- und Grundschulalter* [Mental games with Elfe and Mathis. Fostering reasoning in preschool and primary school age children]. *Manual*. Göttingen, Germany: Hogrefe.

Piazza, M., Facoetti, A., Trussardi, A., Berteletti, I., Conte, S., Lucangeli, D., . . . Zorzi, M. (2010). Developmental trajectory of number acuity reveals a severe impairment in developmental dyscalculia. *Cognition, 116*, 33–41.

Schuchardt, K., Maehler, C., & Hasselhorn, M. (2008). Working memory deficits in children with different learning disabilities. *Journal of Learning Disabilities, 41*, 514–523.

Shepard, R. N., & Metzler, J. (1971). Mental rotation of three-dimensional objects. *Science, 171*, 701–703.

World Health Organization. (2010). *ICD-10: International statistical classification of diseases and related health problems*. New York, NY: Author.

Heinz Holling

Faculty of Psychology and Exercise and Sport Studies
Chair of Statistics and Methods
University of Münster
Fliednerstr. 21
48149 Münster
Germany
Tel. +49 251 83-38485
Fax +49 251 83-39469
E-mail holling@uni-muenster.de

Appendix A

Matrix Notation for the General Linear Model

The regression equation will be written here in an extended form, that is, including the transformations of the original variables. For example, quadratic terms x^2 or interaction terms such as x_1x_2 may be included in the regression equation. Such terms represent transformations of the original explanatory variables. Qualitative variables can also be included in the regression equation after they have been transformed (see below). Therefore, we write the regression equation in a more general form, which includes transformations of the predictors:

$$Y_i = \beta_0 f_0(x_i) + \beta_1 f_1(x_i) + \beta_2 f_2(x_i) + \ldots + \beta_p f_p(x_i) + \varepsilon_i \quad \text{(A1)}$$

or

$$E(Y_i) = \beta_0 f_0(x_i) + \beta_1 f_1(x_i) + \beta_2 f_2(x_i) + \ldots + \beta_p f_p(x_i). \quad \text{(A2)}$$

The regression equation for simple linear regression is then written as

$$E(Y_i) = \beta_0 f_0(x_i) + \beta_1 f_1(x_i), \quad \text{(A3)}$$

where the quantitative predictor x will be transformed by $f_0(x) = 1$ and $f_1(x) = x$. The regression equation including a quadratic term x^2, a further transformation of the original variable x, that is, $f_2(x) = x^2$, is given as:

$$E(Y) = \beta_0 f_0(x) + \beta_1 f_1(x) + \beta_2 f_2(x). \quad \text{(A4)}$$

The ith observation is now given by

$$Y_i = \boldsymbol{f}^{\mathrm{T}}(\boldsymbol{x}_i)\boldsymbol{\beta} + \varepsilon_i \quad \text{(A5)}$$

or

$$E(Y_i) = \boldsymbol{f}^{\mathrm{T}}(\boldsymbol{x}_i)\boldsymbol{\beta} \quad \text{(A6)}$$

with

$\boldsymbol{f} = (f_0, \ldots, f_p)^{\mathrm{T}}$: vector of known functions,
$\boldsymbol{x}_i = (x_{i1}, \ldots, x_{ik})^{\mathrm{T}}$: vector of the independent variables,
$\boldsymbol{\beta} = (\beta_0, \ldots, \beta_p)^{\mathrm{T}}$ the vector of the parameters to be estimated
ε_i: error term.

The regression equation for the whole sample with size n is written as

$$E(\boldsymbol{Y}) = \boldsymbol{F}\boldsymbol{\beta} \quad \text{(A7)}$$

with

$\boldsymbol{Y} = (Y_1, \ldots, Y_n)^{\mathrm{T}}$: n-dimensional vector of the responses,
$\boldsymbol{F} = (\boldsymbol{f}(\boldsymbol{x}_1), \ldots, \boldsymbol{f}(\boldsymbol{x}_n))^{\mathrm{T}}$: $n \times (p + 1)$ design matrix with the regression function $\boldsymbol{f}(\boldsymbol{x}_i)$ evaluated at the value $\boldsymbol{x}_i$ of the independent variables,
$\boldsymbol{\beta}$: $(p + 1)$-dimensional vector of unknown parameters,
$\boldsymbol{\varepsilon} = (\varepsilon_1, \ldots, \varepsilon_n)^{\mathrm{T}}$: n-dimensional vector of the errors.

As for the simple regression model we assume the errors ε_i to be i.i.d.: $\varepsilon_i \sim N(0, \sigma^2)$. Best linear unbiased estimators for $\boldsymbol{\beta}$ will result by the well-known relationship

$$\hat{\boldsymbol{\beta}} = (\boldsymbol{F}^{\mathrm{T}}\boldsymbol{F})^{-1}\boldsymbol{F}^{\mathrm{T}}\boldsymbol{Y}, \quad \text{(A8)}$$

where $\boldsymbol{F}^{\mathrm{T}}\boldsymbol{F}$ is the information matrix which has to be regular.

$\sigma^2\boldsymbol{F}^{\mathrm{T}}\boldsymbol{F}$ is information matrix and $\boldsymbol{M} = \boldsymbol{F}^{\mathrm{T}}\boldsymbol{F}$ the standardized information matrix.

$\boldsymbol{M}$ may also be computed by the outer product $\boldsymbol{f}(\boldsymbol{x})\boldsymbol{f}^{\mathrm{T}}(\boldsymbol{x})$

$$\boldsymbol{M} = \sum_{i=1}^{n} \boldsymbol{f}(\boldsymbol{x}_i)\boldsymbol{f}^{\mathrm{T}}(\boldsymbol{x}_i). \quad \text{(A9)}$$

This relationship is illustrated by a simple example with three units and two variables.

Let $\boldsymbol{F}$ be

$$\boldsymbol{F} = \begin{pmatrix} 1 & 2 \\ 2 & 1 \\ 3 & 4 \end{pmatrix}$$

then

$$\boldsymbol{F}^{\mathrm{T}} = \begin{pmatrix} 1 & 2 & 3 \\ 2 & 1 & 4 \end{pmatrix}$$

and

$$\boldsymbol{F}^{\mathrm{T}}\boldsymbol{F} = \begin{pmatrix} 14 & 16 \\ 16 & 21 \end{pmatrix}$$

The corresponding outer product is

$$\begin{aligned} &\textstyle\sum_{i=1}^{3} \boldsymbol{f}(\boldsymbol{x}_i)\boldsymbol{f}^{\mathrm{T}}(\boldsymbol{x}_i) \\ &= \begin{pmatrix} 1 \\ 2 \end{pmatrix}(1 \;\; 2) + \begin{pmatrix} 2 \\ 1 \end{pmatrix}(2 \;\; 1) + \begin{pmatrix} 3 \\ 4 \end{pmatrix}(3 \;\; 4) \\ &= \begin{pmatrix} 1 & 2 \\ 2 & 4 \end{pmatrix} + \begin{pmatrix} 4 & 2 \\ 2 & 1 \end{pmatrix} + \begin{pmatrix} 9 & 12 \\ 12 & 16 \end{pmatrix} = \begin{pmatrix} 14 & 16 \\ 16 & 21 \end{pmatrix} \end{aligned}$$

The normalized information matrix $\boldsymbol{M}(\xi_n)$ which is adjusted for the sample size n is given for discrete designs with m distinct design points by

$$\boldsymbol{M}(\xi_n) = \frac{1}{n}\boldsymbol{M} = \sum\nolimits_{i=1}^{m} \frac{n_i}{n}\boldsymbol{f}(\boldsymbol{x}_i)\boldsymbol{f}^{\mathrm{T}}(\boldsymbol{x}_i). \quad \text{(A10)}$$

Based on approximate designs the normalized information matrix $\boldsymbol{M}(\xi)$ is computed by

$$\boldsymbol{M}(\xi) = \sum\nolimits_{i=1}^{m} w_i \boldsymbol{f}(\boldsymbol{x}_i)\boldsymbol{f}^{\mathrm{T}}(\boldsymbol{x}_i) \quad \text{(A11)}$$

The predicted response $\widehat{E(Y)} = \hat{Y}(\boldsymbol{x})$ is given by $\boldsymbol{f}^{\mathrm{T}}(\boldsymbol{x})\hat{\boldsymbol{\beta}}$ and its variance by

$$\mathrm{Var}(\hat{Y}(\mathbf{x})) = \sigma^2\boldsymbol{f}^{\mathrm{T}}(\boldsymbol{x})(\mathbf{F}^{\mathrm{T}}\mathbf{F})^{-1}\boldsymbol{f}(\boldsymbol{x}). \quad \text{(A12)}$$

The variance of the predicted response for approximate designs is defined by

$$\frac{\sigma^2}{n}\boldsymbol{f}^{\mathrm{T}}(\boldsymbol{x})\boldsymbol{M}^{-1}(\xi)\boldsymbol{f}(\boldsymbol{x}) \quad \text{(A13)}$$

and for discrete designs by

$$\frac{\sigma^2}{n}\boldsymbol{f}^{\mathrm{T}}(\boldsymbol{x})\boldsymbol{M}^{-1}(\xi_n)\boldsymbol{f}(\boldsymbol{x}). \quad \text{(A14)}$$

The standardized variance which is adjusted for the sample size as well as for the error variance σ^2 is defined for exact designs by

$$\mathrm{d}(\boldsymbol{x}, \xi_n) = \tfrac{n}{\sigma^2}\mathrm{Var}(\hat{Y}(\boldsymbol{x})) = \boldsymbol{f}^{\mathrm{T}}(\boldsymbol{x})\boldsymbol{M}^{-1}(\xi_n)\boldsymbol{f}(\boldsymbol{x}) \qquad (A15)$$

and for approximate designs by

$$\mathrm{d}(\boldsymbol{x}, \xi) = \boldsymbol{f}^{\mathrm{T}}(\boldsymbol{x})\boldsymbol{M}^{-1}(\xi)\boldsymbol{f}(\boldsymbol{x}). \qquad (A16)$$

Appendix B

Coding of Qualitative Variables

Qualitative (categorical) variables have to be transformed by, for example a dummy or effect coding to be included in the regression equation of the general linear model. For instance, if four experimental groups may be represented by a categorical variable x with arbitrary values 1, 2, 3, and 4, this categorical variable has to be transformed by four functions $f_0(x), f_1(x), f_2(x)$, and $f_3(x)$, which will then be included in the regression equation (see below). Designs, however, are described by the original variable x, for example by the values 1, 2, 3, and 4 and their corresponding weights w_i as for example

$$\xi = \begin{Bmatrix} 1 & 2 & 3 & 4 \\ .3 & .2 & .2 & .3 \end{Bmatrix}. \qquad (A17)$$

Given a categorical predictor with r values, this variable will usually be transformed to a constant and $r - 1$ "new" design variables. Some standard transformations exist, such as dummy coding, effect coding, or Helmert contrast coding.

For example, we consider the evaluation of a training for reducing dyscalculia by two experimental groups and two control groups. Let x be a categorical variable describing the group membership with $x = 1$ for a member of E1, $x = 2$ for a member of E2, $x = 3$ for a member of C1, and $x = 4$ for a member of C2.

We now create four "new" design variables by $f_0(x)$, $f_1(x), f_2(x)$, and $f_3(x)$ according to the so-called dummy coding, which is shown in Table 9.

The regression coefficients in the corresponding linear model $E(Y) = \beta_0 f_0(x) + \beta_1 f_1(x) + \beta_2 f_2(x) + \beta_3 f_3(x)$ can be interpreted as follows: β_0 corresponds to μ_{C2}, the population mean of the second control group, which is the so-called reference group here and which is described by the four values $f_0(x) = 1$, $f_1(x) = 0$, $f_1(x) = 0$, $f_2(x) = 0$, and $f_3(x) = 0$. The regression coefficients β_1 belonging to $f_I(x)$ amount to $\mu_{E1} - \mu_{C2}$, the difference of the population means of group E1 and C2, while $\beta_2 = \mu_{E2} - \mu_{C2}$ and $\beta_3 = \mu_{C1} - \mu_{C2}$. Summarizing, the following relationships between regression coefficients and means are given:

$$\begin{aligned} \beta_0 &= \mu_{C2} \\ \beta_1 &= \mu_{E1} - \mu_{C2} \\ \beta_2 &= \mu_{E2} - \mu_{C2} \\ \beta_3 &= \mu_{C1} - \mu_{C2} \end{aligned}. \qquad (A18)$$

Table 9. Dummy coding for a categorical variable with four values and reference group C2

Group	x	$f_0(x)$	$f_1(x)$	$f_2(x)$	$f_3(x)$
E1	1	1	1	0	0
E2	2	1	0	1	0
C1	3	1	0	0	1
C2	4	1	0	0	0

Table 10. Cell mean coding for a categorical variable with four values

Group	x	$f_1(x)$	$f_2(x)$	$f_3(x)$	$f_4(x)$
E1	1	1	0	0	0
E2	2	0	1	0	0
C1	3	0	0	1	0
C2	4	0	0	0	1

Another coding scheme, the so-called cell mean coding, is based on a regression equation without an intercept. For our example with four groups corresponding to a categorical variable with four values, the regression equation now is:

$$E(Y) = \beta_1 f_1(x) + \beta_2 f_2(x) + \beta_3 f_3(x) + \beta_4 f_4(x), \qquad (A19)$$

where $f_i(x)$ are specified as shown in Table 10.

Here, the following relationships between the regression coefficients and the means are obtained

$$\begin{aligned} \beta_1 &= \mu_{E1} \\ \beta_2 &= \mu_{E2} \\ \beta_3 &= \mu_{C1} \\ \beta_4 &= \mu_{C2} \end{aligned}. \qquad (A20)$$

Cell mean coding is especially useful for analyzing linear combinations of means when, for example the two experimental groups are compared to two control groups.

Appendix C

Matrix Notation for Optimality Criteria

D-optimality: Minimizing log det($\boldsymbol{M}^{-1}$) or equivalently $-$log det($\boldsymbol{M}$), with $\boldsymbol{M}$ as standardized information matrix.

A-optimality: Minimizing the trace of $\boldsymbol{M}^{-1}$.

c-optimality: Minimizing the variance of $\boldsymbol{c}^{\mathrm{T}}\hat{\boldsymbol{\beta}}$ corresponds to minimizing $\boldsymbol{c}^{\mathrm{T}}\boldsymbol{M}^{-1}\boldsymbol{c}$.

D_A-optimality: Minimizing log det($\boldsymbol{A}^{\mathrm{T}}\boldsymbol{M}^{-1}\boldsymbol{A}$), with $\boldsymbol{A}$ being the matrix, which contains the coefficients for the linear combinations of the parameters to be estimated.

G-optimality: Determining $\min_{\xi\in\Xi} \max_{\boldsymbol{x}\in\boldsymbol{X}} d(\boldsymbol{x},\xi)$, with $\boldsymbol{X}$ as design region and Ξ as the set of all designs.

IMSE-optimality: Determining $\min_{\xi\in\Xi} \int_{\boldsymbol{X}} d(\boldsymbol{x},\xi)d\boldsymbol{x}$.

Appendix D

c-Optimality for Quadratic Regression Models

For computing the standardized information matrix $\boldsymbol{M} = \boldsymbol{F}^{\mathrm{T}}\boldsymbol{F}$ of the quadratic model $E(Y_i) = \beta_0 + \beta_1 x_i + \beta_2 x_i^2$, with

$$F = \begin{pmatrix} 1 & x_1 & x_1^2 \\ 1 & x_2 & x_2^2 \\ \vdots & \vdots & \vdots \\ 1 & x_n & x_n^2 \end{pmatrix} \tag{A21}$$

We define $S_1 = (\Sigma x_i)/n$, $S_2 = (\Sigma x_i^2)/n$, $S_3 = (\Sigma x_i^3)/n$ and $S_4 = (\Sigma x_i^4)/n$. Then

$$\boldsymbol{F}^{\mathrm{T}}\boldsymbol{F} = \begin{pmatrix} 1 & S_1 & S_2 \\ S_1 & S_2 & S_3 \\ S_2 & S_3 & S_4 \end{pmatrix}. \tag{A22}$$

The inverse $(\boldsymbol{F}^{\mathrm{T}}\boldsymbol{F})^{-1}$ is given by

$$\begin{aligned}(\boldsymbol{F}^{\mathrm{T}}\boldsymbol{F})^{-1} &= 1/\left(n\left(S_2S_4 + 2S_1S_2S_3 - S_2^3 - S_3^2 - S_4S_1^2\right)\right) \\ &\times \begin{pmatrix} S_2S_4 - S_3^2 & S_2S_3 - S_1S_4 & S_1S_3 - S_2^2 \\ S_2S_3 - S_1S_4 & S_4 - S_2^2 & S_1S_2 - S_3 \\ S_1S_3 - S_2^2 & S_1S_2 - S_3 & S_2 - S_1^2 \end{pmatrix}.\end{aligned} \tag{A23}$$

Thus, it follows

$$\begin{aligned}\mathrm{Var}\left(\hat{\beta}_0\right) &= \sigma^2\left(S_2S_4 - S_3^2\right)/ \\ &\left(n\left(S_2S_4 + 2S_1S_2S_3 - S_2^3 - S_3^2 - S_4S_1^2\right)\right),\end{aligned} \tag{A24}$$

$$\begin{aligned}\mathrm{Var}\left(\hat{\beta}_1\right) &= \sigma^2\left(S_4 - S_2^2\right)/ \\ &\left(n\left(S_2S_4 + 2S_1S_2S_3 - S_2^3 - S_3^2 - S_4S_1^2\right)\right),\end{aligned} \tag{A25}$$

$$\begin{aligned}\mathrm{Var}\left(\hat{\beta}_2\right) &= \sigma^2\left(S_2 - S_1^2\right)/ \\ &\left(n\left(S_2S_4 + 2S_1S_2S_3 - S_2^3 - S_3^2 - S_4S_1^2\right)\right).\end{aligned} \tag{A26}$$

Given the design region $[-1, 1]$ the observations have to be allocated in this interval such that the variance of the parameter estimator of interest is minimized. It is obvious that this is a rather complicated optimization problem and approximate optimal design theory should be used instead.

Appendix E

Fedorov Exchange Algorithm

Starting point is a design $\xi_{n(0)}$ with n experimental units and corresponding design matrix $\boldsymbol{F}_{(0)}$ and information matrix $\boldsymbol{F}_{(0)}^{\mathrm{T}}\boldsymbol{F}_{(0)}$. The following steps always consist of two parts, first a further design point is added and then from this extended design one design point is deleted.

Now in the first step a design point $\boldsymbol{x}_i$ is added to the previous design which leads to the new design matrix $\boldsymbol{F}_{(1)}$ with $n + 1$ design points and the corresponding information matrix $\boldsymbol{F}_{(1)}^{\mathrm{T}}\boldsymbol{F}_{(1)}$.

The relation between the "old" and "new" information matrix can be expressed by

$$\boldsymbol{F}_{(1)}^{\mathrm{T}}\boldsymbol{F}_{(1)} = \boldsymbol{F}_{(0)}^{\mathrm{T}}\boldsymbol{F}_{(0)} + \boldsymbol{f}(\boldsymbol{x}_i)\boldsymbol{f}^{\mathrm{T}}(\boldsymbol{x}_i). \tag{A27}$$

The standardized variance $\mathrm{d}(\boldsymbol{x}_i, \xi_{n(0)})$ is given by

$$\mathrm{d}(\boldsymbol{x}_i, \xi_{n(0)}) = \boldsymbol{f}^{\mathrm{T}}(\boldsymbol{x}_i)(\boldsymbol{F}_{(0)}^{\mathrm{T}})\boldsymbol{F}_{(0)})^{-1}\boldsymbol{f}(\boldsymbol{x}_i). \tag{A28}$$

Furthermore, the determinants of the corresponding information matrices are related to each other by

$$\det\left(\boldsymbol{F}_{(1)}^{\mathrm{T}}\boldsymbol{F}_{(1)}\right) = \det\left(\boldsymbol{F}_{(0)}^{\mathrm{T}}\boldsymbol{F}_{(0)}\right)\left(1 + \mathrm{d}(\boldsymbol{x}_{i,}\xi_{n(0)})\right). \tag{A29}$$

The design point $\boldsymbol{x}_i$ is chosen such that its addition maximizes the increase of the determinant $\det(\boldsymbol{F}_{(1)}^{\mathrm{T}}\boldsymbol{F}_{(1)})$ and thus the standardized variance. Then, a design point $\boldsymbol{x}_j$ of the extended design $\xi_{n+1(1)}$ will be deleted according to the following procedure.

The standardized variance function for any design point $\boldsymbol{x}_j$ of the extended design with $\boldsymbol{F}_{(1)}$ is given by

$$\mathrm{d}\left(\boldsymbol{x}_{j,}\xi_{n+1(1)}\right) = \boldsymbol{f}^{\mathrm{T}}(\boldsymbol{x}_j)(\boldsymbol{F}_{(1)}^{\mathrm{T}}\boldsymbol{F}_{(1)})^{-1}\boldsymbol{f}(\boldsymbol{x}_j). \tag{A30}$$

Eliminating any design point $\boldsymbol{x}_j$ leads to a new design matrix $\boldsymbol{F}_{(2)}$ and corresponding information matrix $\boldsymbol{F}_{(2)}^{\mathrm{T}}\boldsymbol{F}_{(2)}$. The matrices $\boldsymbol{F}_{(2)}^{\mathrm{T}}\boldsymbol{F}_{(2)}$ and $\boldsymbol{F}_{(1)}^{\mathrm{T}}\boldsymbol{F}_{(1)}$ are related by

$$\boldsymbol{F}_{(1)}^{\mathrm{T}}\boldsymbol{F}_{(1)} = \boldsymbol{F}_{(2)}^{\mathrm{T}}\boldsymbol{F}_{(2)} - \boldsymbol{f}(\boldsymbol{x}_j)\boldsymbol{f}^{\mathrm{T}}(\boldsymbol{x}_j) \tag{A31}$$

and the determinants by

$$\det\boldsymbol{F}_{(2)}^{\mathrm{T}}\boldsymbol{F}_{(2)} = \det\left(\boldsymbol{F}_{(1)}^{\mathrm{T}}\boldsymbol{F}_{(1)}\right)\left(1 - \mathrm{d}(\boldsymbol{x}_{j,}\xi_{n+1(1)})\right). \tag{A32}$$

Point $\boldsymbol{x}_j$ is chosen such that its elimination leads to the minimum decrease of $\det(\boldsymbol{F}_{(2)}^{\mathrm{T}}\boldsymbol{F}_{(2)})$ and minimizes the standardized variance.

According to this rationale in the next steps a new design point is added and then one design point of the extended design is deleted. The algorithm stops when a certain criterion is fulfilled, for example the increase of the determinant $\det(\boldsymbol{F}_{p+1}^{\mathrm{T}}\,\boldsymbol{F}_{p+1})$ compared to $\det(\boldsymbol{F}_p^{\mathrm{T}}\,\boldsymbol{F}_p)$ is smaller than a defined value.

Original Article

Optimal Experimental Design With Nesting of Persons in Organizations

Gerard J. P. van Breukelen

Faculty of Psychology and Neuroscience, and CAPHRI School for Public Health and Primary Care, Maastricht University, The Netherlands

Abstract. This paper introduces optimal design of randomized experiments where individuals are nested within organizations, such as schools, health centers, or companies. The focus is on nested designs with two levels (organization, individual) and two treatment conditions (treated, control), with treatment assignment to organizations, or to individuals within organizations. For each type of assignment, a multilevel model is first presented for the analysis of a quantitative dependent variable or outcome. Simple equations are then given for the optimal sample size per level (number of organizations, number of individuals) as a function of the sampling cost and outcome variance at each level, with realistic examples. Next, it is explained how the equations can be applied if the dependent variable is dichotomous, or if there are covariates in the model, or if the effects of two treatment factors are studied in a factorial nested design, or if the dependent variable is repeatedly measured. Designs with three levels of nesting and the optimal number of repeated measures are briefly discussed, and the paper ends with a short discussion of robust design.

Keywords: optimal design, multilevel analysis, cost function, sample size, robust design

Multilevel studies are encountered in Psychology in various shapes: Nesting of individuals within organizational units, for instance patients in mental health centers, or pupils in schools. Or nesting of repeated measures within persons, such as follow-ups in a psychotherapy trial or a cohort study in developmental psychology. Both forms of nesting can even be combined into a three-level design, as in longitudinal educational studies with as levels school, pupil, and time. The research question is usually about the effect of predictors like treatment or age on health or behavior. It is then important that the study has sufficient power to detect an effect or association of interest. Power increases with sample size, but resources (time, money, human subjects) are limited, making it important to maximize power given the resources, or to minimize the resources needed for a given power. This is the topic of optimal design.

Early papers on sample size and power (of testing) or precision (of estimation) for multilevel studies are Donner, Birkett, and Buck (1981) and Snijders and Bosker (1993), while Maas and Hox (2005) is a more recent one. McClelland (1997) gives an elementary introduction to optimal design for psychology. Comprehensive introductions are Berger and Wong (2009) for behavioral and biomedical sciences, and Goos and Jones (2011) for industrial applications, while Atkinson, Donev, and Tobias (2007) give a technical account. The combination of optimal design and multilevel studies has been studied for some decades now, see for instance Berger (1986) on longitudinal studies, and Raudenbush (1997) on cluster randomized trials. This paper gives a review of that interesting combination. The outline is as follows: The focus is on optimal design of randomized experiments in a two-level nested setting, like pupils in schools or patients in clinics. Organizations in a nested design can be treated as random or fixed, depending on the number of organizations sampled and on whether one wants to generalize to a population of organizations, among others (Senn, 1998). This paper treats organizations as random. Topics are the optimality criterion for the design, the cost function, the level of randomization, and the optimal sample size per level. First, nested experiments with two conditions and a quantitative dependent variable, such as the total score on a clinical questionnaire or a cognitive test, are discussed. After that, covariates, dichotomous outcomes, factorial nested designs, and nested designs with two or three repeated measures will be covered, followed by a brief discussion of designs with three levels of nesting and designs with more than three repeated measures, and ending with some discussion of robust design.

Randomized Experiments with Nesting of Persons in Organizations

Randomized experiments to evaluate a treatment are often run in nested settings. For example, Ausems, Mesters, Van Breukelen, and De Vries (2002) evaluated two smoking prevention programs, and Kraag, Van Breukelen,

DOI: 10.1027/2151-2604/a000143

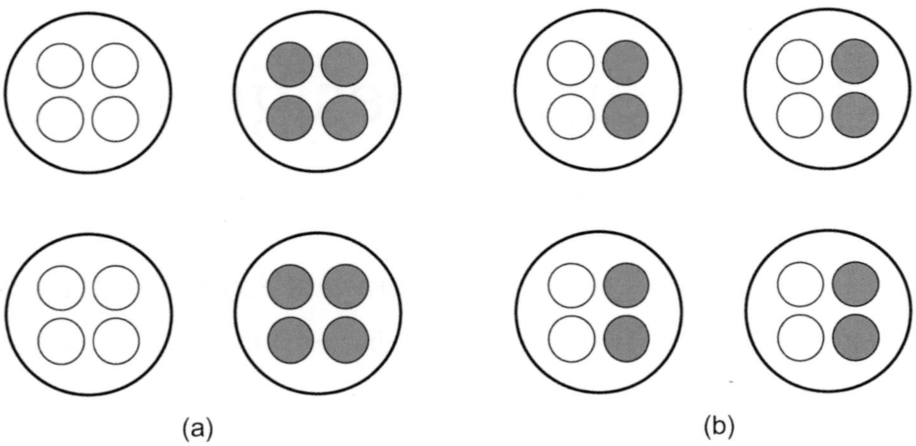

Figure 1. Design of (a) cluster randomized trial and (b) multisite trial, with four clusters or sites and four persons per cluster or site (solid = treated, clear = control).

Kok, and Hosman (2009) tested a stress management program, for children in primary schools. Similarly, Van Keulen et al. (2011) tested two interventions to promote healthy behaviors (physical activity, healthy food) among patients of general practices. Given this nesting, randomization to treatment conditions can be done at either design level: organization or individual. The first type is known as the cluster randomized trial (Donner & Klar, 2000; Eldridge & Kerry, 2012; Hayes & Moulton, 2009; Raudenbush, 1997), or as the group randomized trial (Murray, 1998). The second is known as the multisite or multicenter trial (Raudenbush & Liu, 2000; Senn, 1998). The term "trial" is used in medical and health research to denote a randomized experiment. The difference between a cluster randomized and multisite trial is shown in Figure 1, and is similar to the difference between a between-subject and within-subject design in psychology, but now with organizations as subjects, and subjects as replications or trials to be averaged into one outcome per organization per treatment. A cluster randomized trial is a between-organization design which can be analyzed with the unpaired *t*-test, using the outcome mean per organization as dependent variable. A multisite trial is a within-organization design which can be analyzed with the paired *t*-test, using the outcome mean per treatment per organization as dependent variable. Both designs are usually analyzed with multilevel regression of the individual data, but this is equivalent to the *t*-tests above if the sample size is the same in all organizations and the outcome is quantitative and there are no covariates (Moerbeek, Van Breukelen, & Berger, 2000). If the sample size varies or if covariates are involved, then multilevel analysis is needed.

Many psychologists will know that a within-subject design is more powerful, thus requiring a smaller sample size, than a between-subject design. In the between-subject design, all outcome variation between individuals in the same condition enters into the sampling variance of the treatment effect estimator, as the standard error of the unpaired *t*-test and the denominator of the ANOVA *F*-test show. In the within-subject design, two (or more) treatments are compared within the same individual and so all interindividual variation drops out of the treatment effect, leaving only intraindividual variation due to person by treatment interaction and measurement error as sources of sampling error. An analogous result holds for cluster randomized trials versus multisite trials, now with organizations instead of individuals. Multisite trials are thus more powerful and require fewer organizations than cluster randomized trials. In spite of this, cluster randomized trials are common, either for logistic reasons or to prevent treatment contamination. To evaluate a new teaching method in classroom one can only randomly assign classes, not pupils. If class randomization leads to treatment contamination due to communication between treated and untreated classes or between their teachers, one may even need to randomize schools instead of classes. After all, treatment contamination leads to underestimation of the treatment effect and a loss of power (Moerbeek, 2005; Rhoads, 2011). This is analogous to the problem of carry-over effects in within-subject designs.

The consideration that multisite trials have more power, but are also more prone to treatment contamination, than cluster randomized trials, has even led to the recent development of three hybrids, the pseudo-cluster randomized trial (Borm, Melis, Teerenstra, & Peer, 2005; Teerenstra, Moerbeek, Melis, & Borm, 2007), the cluster randomized crossover trial (Giradeau, Ravaud, & Donner, 2008; Rietbergen & Moerbeek, 2011; Turner, White, & Croudache, 2007), and the stepped wedge cluster randomized trial (Hayes & Moulton, 2009; Hussey & Hughes, 2007). Further, experiments with clustering in one treatment condition only are also encountered, for instance, in comparing individual therapy with group therapy. If patients are nested within groups, their outcomes may be correlated due to shared environmental influences like the therapist and group dynamics (Candel & Van Breukelen, 2009; Moerbeek & Wong, 2008; Roberts & Roberts, 2005). Neither the hybrids nor the case of clustering in one condition will be discussed in this paper, however. Let us start with the cluster randomized trial, which is statistically easier than the multisite trial.

Optimal Design of Cluster Randomized Trials

Kraag et al. (2009) examined the effects of a stress management program on stress, coping, anxiety, and depression in primary school children. They randomly assigned 52 schools to treatment or control, with an average sample of 28 children per school, roughly the class size in the Netherlands. Here, the class level is ignored and is in fact indistinguishable from the school level if only one class participates per school. The question now is how powerful their sample size is and whether it is optimal. To find this out, some equations are needed, starting with a simple model for data analysis:

$$y_{ij} = \beta_{0j} + \beta_1 x_j + \varepsilon_{ij}, \tag{1a}$$

where

$$\beta_{0j} = \beta_0 + u_{0j}. \tag{1b}$$

Here, y_{ij} is the outcome for pupil i in school j, and x_j is the treatment condition to which school j was assigned (1 = treatment, 0 = control, or centered as +0.5 for treatment and −0.5 for control). Parameter β_1 is the outcome difference between treated and control, that is, the treatment effect. The intercept β_{0j} varies randomly between schools, as it is the sum of a fixed mean intercept β_0 plus a random deviation u_{0j} to allow differences between schools. This random effect u_{0j} has mean zero and variance σ_{u0}^2. Finally, ε_{ij} reflects person and measurement error effects and has mean zero and variance σ_ε^2 within each school, implying independence between u_{0j} and ε_{ij}. Due to (1b), this model is known as "random intercept model." Inserting (1b) into (1a) gives our model equation for a cluster randomized trial:

$$y_{ij} = \beta_0 + \beta_1 x_j + u_{0j} + \varepsilon_{ij}, \tag{1c}$$

The outcomes y_{ij} and $y_{i'j}$ of two pupils i and i' in the same school j are correlated due to the common school effect u_{0j}, and this so-called intraclass correlation (ICC) equals:

$$\rho = \frac{\sigma_{u0}^2}{\sigma_y^2} = \frac{\sigma_{u0}^2}{\sigma_{u0}^2 + \sigma_\varepsilon^2}, \tag{2}$$

where the denominator σ_y^2 is the total outcome variance within conditions.

The present model has four unknown parameters, that is, the fixed effects β_0 and β_1, and the variance components σ_{u0}^2 and σ_ε^2. The Maximum Likelihood (ML) estimators of the fixed effects are uncorrelated with those of the variance components (Ankenman, Aviles, & Pinheiro, 2003; Searle, Casella, & McCulloch, 1992). So optimal design criteria may be based on the covariance matrix of the fixed effect estimators,

$$\mathrm{Cov}\begin{pmatrix}\hat\beta_0\\ \hat\beta_1\end{pmatrix} = \begin{pmatrix}\mathrm{Var}(\hat\beta_0) & \mathrm{Cov}(\hat\beta_0\,\hat\beta_1)\\ \mathrm{Cov}(\hat\beta_0\,\hat\beta_1) & \mathrm{Var}(\hat\beta_1)\end{pmatrix}$$

or on the covariance matrix of the variance component estimators

$$\mathrm{Cov}\begin{pmatrix}\hat\sigma_e^2\\ \hat\sigma_{u0}^2\end{pmatrix} = \begin{pmatrix}\mathrm{Var}(\hat\sigma_e^2) & \mathrm{Cov}(\hat\sigma_e^2\,\hat\sigma_{u0}^2)\\ \mathrm{Cov}(\hat\sigma_e^2\,\hat\sigma_{u0}^2) & \mathrm{Var}(\hat\sigma_{u0}^2)\end{pmatrix},$$

or on the total 4-by-4 covariance matrix of all four parameters estimators. The two matrices above contain all information about the precision of the estimated fixed effects and estimated variance components for a given design, respectively. In particular, one may want to design the study such that it minimizes either the trace (A-optimality) or the determinant (D-optimality) of either covariance matrix. In this paper we focus on the fixed effects, and their covariance matrix is derived in Appendix A. For the covariance matrix of the variance components, see Van Breukelen, Candel, and Berger (2008).

Now, the A-optimality criterion means minimizing the sum, and thereby also the average, of the two variances on the diagonal. A drawback of this criterion is that the A-optimal design in general depends on the scaling of the predictor, for instance, on whether we code treatment x_j in Equation 1a as 0 for control and 1 for treated, or centered as −0.5 for control and +0.5 for treated. In contrast, the D-criterion, which means minimizing the determinant of the covariance matrix, is invariant to such coding, that is, it gives the same optimal design for any rescaling of the predictors. Another nice property of D-optimality is that it minimizes the volume of the confidence ellipsoid for the parameters concerned. On the other hand, a potential drawback of D is that minimizing the volume of a confidence ellipsoid can sometimes lead to a narrow confidence interval for one parameter and a very wide one for another parameter.

For the fixed effects in the present model, it turns out that the A-criterion and D-criterion are both proportional to $\mathrm{Var}(\hat\beta_1)$, the variance of the treatment effect estimator (for the proof, see Appendix A). This is nice because the aim of a cluster randomized trial is usually to estimate the treatment effect β_1 as precisely as possible and to have maximum power for testing this effect. This requires the minimization of the standard error of $\hat\beta_1$, or equivalently, of the squared standard error $\mathrm{Var}(\hat\beta_1)$, which will therefore be the optimality criterion in the sequel. For the present model it can be rewritten as (see Appendix A):

$$\mathrm{Var}(\hat\beta_1) = [(n-1)\rho + 1] \times \frac{4\sigma_y^2}{nK}. \tag{3}$$

Here, n is the common sample size per school and K is the number of schools, and so nK is the total number of pupils sampled. The factor $[(n - 1)\rho + 1]$ is called the design effect (DE) and indicates the sample size needed for a cluster randomized trial relative to that for a classical trial with randomization of individuals. If the ICC ρ is zero, there is no outcome variation between schools, only between pupils within schools. The design effect is 1 then, and cluster randomization is then as efficient as individual randomization. In practice, the ICC is usually between 0.01

and 0.10, giving a design effect larger than 1, which increases with the sample size per school, n. As n gets very large, $\mathrm{Var}(\hat{\beta}_1)$ becomes roughly $\frac{4\rho\sigma_y^2}{K}$, that is, four times the outcome variance at the school level divided by the number of schools sampled. This shows that the number of clusters is more important than the number of subjects per school for the power of a cluster randomized trial.

Now, increasing sample size not only increases power, but also total study cost (money, time, and effort), raising the question of the best sample size per level (school, pupil) of a cluster randomized trial. To answer this question an optimality criterion like (3) is needed, plus a cost function relating study costs to sample size. A simple cost function assumes a fixed cost c per cluster and cost s per subject, where c and s are in the same currency (e.g., Euro). The total budget B needed for a cluster randomized trial with K schools of n pupils each, excluding all costs that do not depend on sample size, then becomes:

$$B = cK + snK = K(c + sn). \quad (4)$$

Note that $(c + sn)$ is the total cost per school including the costs per pupil. The optimal design minimizes $\mathrm{Var}(\hat{\beta}_1)$ in (3) as a function of n and K, within the constraint in (4). The optimum turns out to be (Cohen, 1998; Moerbeek et al., 2000; Raudenbush, 1997; present Appendix A):

$$n = \sqrt{\left(\frac{1-\rho}{\rho}\right)\left(\frac{c}{s}\right)}, \quad K = \frac{B}{(c + sn)}. \quad (5)$$

So the optimal number of pupils sampled per school, n, increases as the school-to-pupil cost ratio c/s increases, or as the pupil-to-school variance ratio $\left(\frac{1-\rho}{\rho}\right)$ increases (i.e., as the ICC decreases). The optimal number of schools sampled, K, then follows from (4). Figure 2 illustrates the effect of an increasing ICC on the optimal design for a given budget and costs. Note that in (5), the budget B only affects the optimal number of schools K, not the optimal number of pupils per school n. So if the budget is increased to gain power, then the extra budget must be spent on sampling more schools. Also, note that the optimal design depends on the ICC, which is usually unknown in the design stage. So the design is locally optimal only, that is, optimal for one specific value of the unknown ICC. In contrast, it does not depend on the unknown fixed effects β_0 and β_1, and so, in this respect, it is globally optimal.

Combining (3) and (5) gives $\mathrm{Var}(\hat{\beta}_1)$ for the optimal design. This is the smallest possible $\mathrm{Var}(\hat{\beta}_1)$, thus giving the largest possible power and precision within our budget, given the sampling costs, outcome variance, and ICC. Depending on the effect size of interest, it may be that this power is still insufficient. The only solution then is to increase the budget B and the number of clusters K, using the following equation for relating the sample size to the effect size and power:

$$K = 4\left(\frac{\mathrm{DE}}{n}\right) \times \left(z_{1-\gamma} + z_{1-\alpha/2}\right)^2 \times \left(\frac{1}{d}\right)^2. \quad (6)$$

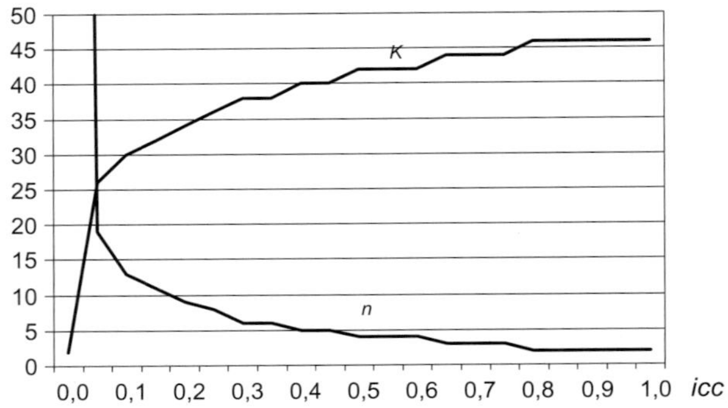

Figure 2. Optimal number of clusters K and number of persons per cluster n as a function of the ICC (budget B = 100.000, cost per cluster c = 2,000, cost per subject s = 100). Numbers are rounded to the nearest integer, K is constrained to be even to allow 50:50 allocation, and K and n are both constrained to be at least 2.

Here, DE is the design effect $[(n - 1)\rho + 1]$, and $z_{1-\gamma}$ is the $100(1 - \gamma)$th percentile of the standard normal distribution (i.e., 1.28 for a power of 90%, or 0.84 for a power of 80%), and $z_{1-\alpha/2}$ is the $100(1 - \alpha/2)$th percentile (i.e., 1.96 if α = 0.05 two-tailed). Finally, d is Cohen's standardized effect size $d = \frac{\mu_1 - \mu_2}{\sigma_y}$ which, for model (1) is $d = \frac{\beta_1}{\sigma_y}$, since μ_1 and μ_2 are the expected outcomes in the two respective treatment conditions, that is, the fixed part of Equation 1c. By substituting (5) for n in (6), the minimum K and budget needed are found, given the effect size, power, and α. If ρ = 0, the DE = 1 and (6) reduces to the sample size equation for the classical between-subject experiment after multiplying both sides of (6) with n. A simple derivation of the sample size Equation 6 is given in Appendix B.

Example

Let us apply the theory to the stress management trial in primary school, which included K = 52 schools with a sample of n = 28 children per school (Kraag et al., 2009). Let us first consider what the optimal design would be and how efficient the actual design was, and then discuss some practical complications and solutions.

To compute the optimal design with (5) one needs to know the school-to-pupil cost ratio c/s and the ICC ρ. Kraag et al. report ICC values between 0.003 and 0.03, depending on the outcome variable, so let us take ρ = 0.05 to be on the safe side. Since the authors do not mention study costs, assume a cost ratio c/s = 25. Equation 5 then gives as optimal sample size per school n = 22, somewhat less than the actual n = 28. It follows from (5) and the cost ratio that the optimal design can have 12% more schools than the actual design for the same budget: K = 58 instead of 52. Finally, with (3) we can check that the ratio of $\mathrm{Var}(\hat{\beta}_1)$ for the optimal design (n = 22, K = 58) and $\mathrm{Var}(\hat{\beta}_1)$ for the actual design (n = 28, K = 52) is close to 1, since the design effects DE are 2.05 optimal and 2.35 actual, and the sample sizes nK are 1276 optimal and 1456 actual.

So, while the actual design was not optimal, it was nearly so, for $\rho = 0.05$ and $c/s = 25$.

Of course, another cost ratio or ICC gives a different optimum. For instance, an ICC of 0.10 and a cost ratio of 16 give $n = 12$ and $K = 82$ for the same budget B as for the actual design. Following (3) and (5) this gives a DE of 2.1 optimal versus 3.7 actual, a total sample size $nK = 984$ optimal versus 1,456 actual, and a ratio of actual to optimal Var $(\hat{\beta}_1)$ of 1.19. So the actual design now has a 19% larger sampling variance than the optimal design. It therefore needs a 19% larger budget and 19% extra schools (i.e., $K = 62$ instead of 52 schools) to have the same power as the optimal design.

But what effect size can be detected with a power of, say, 90%, with the actual design? As Equation 6 shows, this not only depends on n and K, but also on the ICC and the α used for significance testing. Filling in a typical ICC of 0.05, and $\alpha = 0.05$ two-tailed, a power $(1 - \gamma) = 0.90$, and sample size $n = 28$ and $K = 52$, gives $d = 0.26$ which, according to Cohen's classification, is a small effect. For a medium effect $d = 0.50$, the power of the actual design can be verified to be almost 100%. So the actual design has quite some power to detect a treatment effect. Again, this depends on the assumed ICC of course. If the ICC were 0.10, the actual design would have a power of 90% for an effect size $d = 0.33$, which is still a small to medium effect size. Even taking into account that Kraag et al. used $\alpha = 0.01$ to correct for multiple outcome testing, the actual design would have 90% power for an effect $d = 0.31$ if ICC $= 0.05$, or for $d = 0.39$ if ICC $= 0.10$. The largest ICC found by Kraag et al. was only 0.03, and so no worry is needed about the power of their trial for clinically meaningful effect sizes.

In practice, some complications occur and here a rule of thumb is given for handling each, referring to Van Breukelen, Candel, and Berger (2007) and to Van Breukelen and Candel (2011, 2012) for more details. The first complication is that the optimal n or optimal K may not be feasible. One can then take n or K as close to the optimum as possible and adapt the other factor, K or n, to have enough power. Increasing K to compensate a small n is more effective than increasing n to compensate a small K, however, as increasing n increases the design effect, see (3). Another complication is that the outcome variance and ICC are unknown, but needed to compute the optimal design and budget needed, see (5) and (6). This can be solved by taking as ICC the midpoint of its plausible range (e.g., 0.05 for a range from 0 to 0.10), and increasing K with 2 if $\alpha = 0.05$, or with 4 if $\alpha = 0.01$, two-tailed. A last complication is that the sample size n usually varies between clusters, due to differences in natural cluster size and dropout. This leads to a loss of power, which can usually be compensated by increasing the number of clusters with about 10%.

Optimal Design of Multisite/ Multicenter Trials

As explained in the Optimal Design of Cluster Randomized Trials section, multisite or multicenter trials (Raudenbush & Liu, 2000; Senn, 1998) are to cluster randomized trials what within-subject designs are to between-subject designs. In a multisite trial there is random assignment of individuals within each site (organization, cluster, center) to treatment or control, and so treatment varies within instead of between sites, see Figure 1. This gives more power and precision than cluster randomization, at the price of potential treatment contamination due to communication between individuals within sites. A multisite trial is therefore the design of choice when individual randomization is feasible, treatment contamination is ignorable, and multiple sites are needed either to include a sufficient number of individuals or simply because one wants to test treatment by site interaction. As indicated before, site can be treated as fixed or random, depending on various considerations, but as a rule of thumb, one may treat site as random in trials with at least 10 sites and as fixed otherwise. The reason for this is that having fewer than 10 sites may prevent estimation of random site effects, and having more than 10 sites may give too many unknown regression parameters if site is treated as fixed. Note, however, that for precise estimation with site as random 20 rather than 10 sites may be advisable (Snijders & Bosker, 2011).

An example of a multisite trial is the Vitalum trial of Van Keulen et al. (2011), which tested two different interventions to promote healthy behaviors (physical activity, healthy food) among patients of general practices. In each of 23 practices, patients meeting the inclusion criteria were randomly assigned to one of four treatment conditions (intervention A, intervention B, both interventions, no intervention). The total sample size was 1,629 patients, giving an average of $n = 70.8$ per site. This section assumes that the trial had two conditions only, treatment and control, with 35 patients per condition per site. The actual twoway nested design is covered in a later section.

Analogously to model (1) for a cluster randomized trial with two conditions and a quantitative outcome, one can specify a model for a multisite trial:

$$y_{ij} = \beta_{0j} + \beta_{1j}x_{ij} + \varepsilon_{ij}, \tag{7a}$$

with:

$$\beta_{0j} = \beta_0 + u_{0j}, \tag{7b}$$

and

$$\beta_{1j} = \beta_1 + u_{1j}. \tag{7c}$$

Inserting (7b) and (7c) in (7a) gives:

$$y_{ij} = \beta_0 + \beta_1 x_{ij} + u_{0j} + u_{1j}x_{ij} + \varepsilon_{ij}. \tag{7d}$$

Here, y_{ij} is the outcome for patient i in practice j, and x_{ij} is the treatment which can be coded as 1 for treated and 0 for control, or centered as +0.5 and −0.5. Comparing with model (1) for cluster randomized trials shows that, in addition to the intercept β_{0j}, the slope β_{1j} now also depends on the cluster/center j. It is the sum of a fixed mean slope β_1 plus a random effect u_{1j}, which reflects differences between site with respect to the treatment effect, that is, treatment by site interaction. This model is known as the "random slope model," to distinguish it from the "random intercept model" (1).

Analogously to the random site effect u_{0j}, the random interaction effect u_{1j} has mean zero and variance σ_{u1}^2. Assuming centered coding of x (−0.5/+0.5) and absence of slope-intercept covariance gives as total outcome variance:

$$\sigma_y^2 = \sigma_{u0}^2 + \frac{1}{4}\sigma_{u1}^2 + \sigma_\varepsilon^2, \tag{8}$$

which is homogeneous between treatment conditions. As a result, 50:50 allocation in each site is optimal, and Equation 3 for the variance of the treatment effect, $\mathrm{Var}(\hat{\beta}_1)$, still applies, except that we need to replace σ_{u0}^2 with $\frac{1}{4}\sigma_{u1}^2$ in the definitions of ρ and σ_y^2 in (2) and (3) (for technical details, see Appendix C). Since the treatment effect is inferred from the outcome difference between treated and controls within the same site, the site effect u_{0j} drops out of the treatment effect estimator $\hat{\beta}_1$, and so σ_{u0}^2 has no impact on $\mathrm{Var}(\hat{\beta}_1)$ (Van Breukelen et al., 2007). From this, it follows that Equations 5 and 6 can be used for multisite trials by replacing σ_{u0}^2 with $\frac{1}{4}\sigma_{u1}^2$ (see Appendix C). Note that the between-cluster variance σ_{u0}^2 in model (1) is actually the sum of the site and site by treatment variance components in (8), since these two components are not separable in case of cluster randomization. This is analogous to a between-subject design, where person by condition interaction is not separable from residual (person) effects. As a result, $\mathrm{Var}(\hat{\beta}_1)$ is always larger for a cluster randomized trial than for a multisite trial, making the latter more powerful, unless treatment contamination occurs (Moerbeek, 2005; Rhoads, 2011).

What does this theory imply for the lifestyle trial of Van Keulen et al. (2011)? The trial design was based on a power calculation which assumed a small effect $d = 0.3$, a power of 90%, a two-tailed $\alpha = 0.01$ to correct for multiple testing, an average of $n = 70$ included patients per general practice within the study time frame, and an ICC of 0.02 (Van Keulen et al., 2008). A small ICC was assumed because only treatment by site effects, not site effects, affects $\mathrm{Var}(\hat{\beta}_1)$. Filling in these numbers in (6) gives $K = 23$ practices, and so the design was sufficiently powered. But was it optimal? This depends on the practice-to-patient cost ratio. Given the assumed ICC of 0.02, it follows from (5) that the actual $n = 70$ per site is optimal for a cost ratio $c/s = 100$. In contrast, if $c/s = 25$, the optimal sample size per site is $n = 35$ and according to (5) one can then have $K = 36$ instead of 23 practices for the same price. Fortunately, it may be verified with (3) that the actual design is still fairly efficient for $c/s = 25$, giving a $\mathrm{Var}(\hat{\beta}_1)$ which is only 11% larger than for the optimal design. So the actual design would need a 11% larger budget than the optimal one to have the same power, and this extra budget should then be spent on a few extra sites, that is, K becomes 25 or 26 instead of 23.

As explained at the start of this section, the lifestyle trial actually had a twoway design, since patients in each center were randomly assigned to one of four treatment conditions (treatment A, treatment B, A and B, none). A later section will discuss the twoway nested design and show how the methods can be applied to that design.

Covariates

The models (1) for cluster randomized trials and (7) for multisite trials do not include any covariates, but in practice covariates are usually included into the data analysis. In nonrandomized studies this primarily serves to correct for confounding. In randomized experiments it mainly serves to increase power and precision. We will first see how this works for the classical between-subject experiment, and then generalize that to nested designs. Denote the within-group correlation between covariate C and outcome Y as $\rho_{yc.x}$, where X indicates the treatment group. Including C into the analysis then gives an unexplained outcome variance which is $(1 - \rho_{yc.x}^2)$ as large as without C (Rausch, Maxwell, & Kelley, 2003; Senn, 1989). This unexplained variance is the denominator of the squared effect size, d^2 which thus increases, leading to a sample size which needs to be only $(1 - \rho_{yc.x}^2)$ as large as without C. The most important covariate is usually a pretest of the outcome itself and correlations of 0.50 up to 0.70 with the posttest are quite usual. So by including a pretest as covariate we can reduce the sample size with 25% to 50%, depending on the pre-post correlation. This only holds for randomized experiments, since randomization ensures that treatment and covariate are not correlated with each other apart from sampling error (i.e., $\rho_{xc} = 0$). In nonrandomized studies, there will be such correlation and covariate adjustment may then increase or decrease the variance of the treatment effect estimator, depending on whether ρ_{xc} or $\rho_{yc.x}$ is larger (for details, see, e.g., Fox, 1997, p. 338). However, in nonrandomized studies covariates serve to correct for confounding rather than to improve power, and this correction entails complications (Van Breukelen, 2006).

But what does this mean for covariate adjustment in cluster randomized and multisite trials? Work on this question was done by Moerbeek, Van Breukelen, and Berger (2001a, 2008) and can be summarized as follows. Including a cluster-level covariate into (1) or (7), for instance school type, reduces cluster-level variance σ_{u0}^2, at least if the covariate is predictive of the outcome. Likewise, including a person-level covariate like pupil's age reduces person-level variance σ_ε^2 (Moerbeek et al., 2008, p. 188). Both effects improve precision and power for the treatment effect, see (3). On the other hand, each covariate costs a degree of freedom. Given that the *df* for the treatment effect is $K - 2$ in a cluster randomized trial, or $K - 1$ in a multicenter trial with random slope (Moerbeek, Van Breukelen, & Berger, 2003), this loss of *df* due to covariates can only be ignored for large K and small number of covariates. Further, chance correlation between treatment and covariate increases $\mathrm{Var}(\hat{\beta}_1)$ as already explained for non-nested experiments. Such correlation can arise between treatment and a cluster-level covariate in a cluster randomized trial with small K, which has led to the proposal of matching in cluster randomized trials (Hayes & Moulton, 2009; Imai, King, & Nall, 2009; Klar & Donner, 1997). In contrast, treatment-covariate correlation is ignorable for a person-level covariate in a multisite trial if nK is large, and is even zero for a cluster-level covariate in a multisite trial, as well as for a person-level covariate in a cluster randomized trial. A covariate which varies both within and between clusters can be split into a cluster-level part (the cluster mean of the covariate) and a person-level part (the personal deviation from the cluster mean), and so the theory applies to such covariates as well. For further details, see Moerbeek (2006) and Teerenstra, Eldridge, Graff, DeHoop, and Borm (2012).

Dichotomous Outcomes

So far, a quantitative outcome was assumed, like the total score on a stress questionnaire, or the number of hours of physical activity per week. Categorical outcomes are also encountered, however, especially dichotomous ones like recovery from disease or passing an examination. Some cluster randomized trials indeed have a binary outcome, such as being a smoker or not at posttest, as in the smoking prevention trial by Ausems et al. (2002). Likewise, the multisite trial of Van Keulen et al. (2008) included dichotomous outcomes like meeting an established norm for physical activity. Instead of the multilevel (or mixed) linear models (1) and (7) multilevel (or mixed) logistic models are now needed to relate the logodds of smoking or meeting the activity norm, to treatment and cluster effects. The logodds is defined as $\mathrm{LN}\left(\frac{P}{1-P}\right)$, where P is the probability of being a smoker or meeting the norm, respectively.

So, for a dichotomous outcome in a cluster randomized trial, the random intercept model (1) becomes:

$$\mathrm{LN}\left(\frac{P_{ij}}{1-P_{ij}}\right) = \beta_{0j} + \beta_1 x_j = \beta_0 + \beta_1 x_j + u_{0j}. \quad (9)$$

Likewise, for a dichotomous outcome in a multisite trial, the random slope model (7) becomes:

$$\begin{aligned}\mathrm{LN}\left(\frac{P_{ij}}{1-P_{ij}}\right) &= \beta_{0j} + \beta_{1j} x_{ij} \\ &= \beta_0 + \beta_1 x_{ij} + u_{0j} + u_{1j} x_{ij}. \quad (10)\end{aligned}$$

There are no closed form equations for the ML estimators of the model parameters or their covariance matrix, except for so-called first-order Marginal Quasi-Likelihood (MQL1) estimation, a method which is strongly biased, however. Mathematical derivations for this method and simulations for so-called second-order Penalized Quasi-Likelihood (PQL2), a method which is fairly unbiased, suggest that the optimal design results for the mixed linear models (1) and (7) also apply to the mixed logistic models (9) and (10), with two modifications (Candel & Van Breukelen, 2010; Moerbeek, Van Breukelen, & Berger, 2001b). First, in Equations 2, 3, 8, the outcome variance at the individual level, σ_ε^2, which is part of the total outcome variance σ_y^2, must be replaced with the following term δ^2:

$$\delta^2 = \frac{1}{2}\left(\frac{1}{\pi_c(1-\pi_c)} + \frac{1}{\pi_t(1-\pi_t)}\right), \quad (11)$$

where π_c and π_t are the probability P of being a smoker in the control and treated group, respectively, given a random effect of zero (i.e., $u_{0j} = 0$ for a cluster randomized trial, and $u_{1j} = 0$ for a multisite trial). Note that, unlike σ_ε^2 in the mixed linear models (1) and (7), the term δ^2 depends on the fixed effects β_0 and β_1, with a minimum of 4 if $\beta_0 = 0$ and $\beta_1 = 0$ so that π_c and π_t are both 0.50, and a very large value as either π_c or π_t approaches 0 or 1. Consequently, as Equations 2 and 5 show, the ICC and the optimal design for mixed logistic regression depend on the unknown fixed effects of interest. Solutions to this local optimality problem are discussed in the last section of this paper, with some useful references.

A second difference with the mixed linear model is that, for mixed logistic regression, the number of clusters as computed with the equations must be increased with 10% up to 25%. This is because the equations are based on MQL1 parameter estimation which is strongly biased, and simulations showed that PQL2 estimation, which is fairly unbiased but not amenable to mathematical derivations, gives a $\mathrm{Var}(\hat{\beta}_1)$ which is 10% to 25% larger, depending on ρ, n, and K (Candel & Van Breukelen, 2010; Moerbeek et al., 2001b). Apart from this increase with 10% up to 25%, dichotomous outcomes usually need a larger sample than quantitative outcomes also in non-nested settings, due to a smaller effect size. For example, dichotomizing a quantitative outcome on its median reduces the power of a between-subject experiment such that the sample size must be 60% larger. A similar result holds for the sign test relative to the paired t-test for a within-subject design. Dichotomizing a quantitative outcome is thus bad practice from a statistical viewpoint (Fedorov, Mannino, & Zhang, 2009; MacCallum, Zhang, Preacher, & Rucker, 2002; Senn & Julious, 2009). It should only be considered if there are strong substantive reasons for doing so, and in certain special cases (DeCoster, Iselin, & Gallucci, 2009; Vargha, Rudas, Delaney, & Maxwell, 1996).

Finally, the effect of covariates in a mixed logistic model is more complicated than, but probably similar to, the effect in mixed linear models. For the case of fixed logistic regression in the absence of nesting, adding a predictive covariate which is uncorrelated with the treatment indicator X (as in a randomized experiment) increases both the regression weight of X and its standard error, but the net result is a power gain (Hernandez, Steyerberg, & Habbema, 2004; Moore, Neugebauer, Valappil, & Van der Laan, 2011; Robinson & Jewell, 1991). In nonrandomized studies the increasing effect of treatment-covariate correlation ρ_{xc} on the standard error of the effect of X appears to be the same for logistic as for linear regression (Hsieh, Bloch, & Larsen, 1998). However, to the author's knowledge, no

Table 1. Two way factorial ANOVA in terms of three orthogonal contrasts. Each contrast or effect is the weighted sum of the four outcome means, one per condition, with as weights the contrast coefficients in the table. Coefficients per contrast are normalized to have mean zero and sum of squares one

Treatment condition	Main effect of factor *A*	Main effect of factor *B*	Interaction effect of *A* and *B*
Neither *A* nor *B*	−0.5	−0.5	+0.5
A only	+0.5	−0.5	−0.5
B only	−0.5	+0.5	−0.5
Both *A* and *B*	+0.5	+0.5	+0.5

extension of these results to mixed logistic regression has yet been published.

Nested Designs With Two Treatment Factors in a 2 × 2 Factorial Design

Twoway factorial designs study the combined effects of two treatments. Typically, each factor has two levels, for instance medication yes/no and psychotherapy (cognitive/interpersonal) in a recent study of treatments of depression (Peeters et al., 2013). Combining such a twoway design with nesting gives three possible design types:

(1) the fully cluster randomized design where each cluster is randomly assigned to one of the four conditions, so that both factors are cluster-level predictors;
(2) the fully person randomized design where, within each cluster, persons are randomly assigned to conditions, so that both factors are person-level predictors;
(3) a design where clusters are randomly assigned to levels of the first factor, and persons within each cluster are randomly assigned to levels of the second factor, so that one factor varies between clusters and the other within clusters.

An example of the fully cluster randomized 2 × 2 design is the smoking prevention trial of Ausems et al. (2002). Treatment factors were an in-school and an out-of-school intervention and schools were randomly assigned to one of the four combinations. An example of the fully person randomized 2 × 2 design is the Vitalum study on healthy behaviors of Van Keulen et al. (2011). Treatment factors were a telephonic motivational interview and tailored printed communication, and patients within general practices were randomly assigned to one of the four combinations.

Optimal design for these study types is similar to that for trials with two conditions. Usually one will test the interaction between both factors, followed by main effects if there is no interaction, or simple effects if there is interaction. Assuming homogeneity of variance and costs between conditions, it is optimal to have the same number of clusters and the same number of persons in every condition. With such a balanced design, and assuming centered coding of both treatment factors as in ANOVA, the interaction and main effects are three orthogonal contrasts, summarized in Table 1. Every contrast comes down to a comparison between two outcome means, bringing us back to a nested design with two treatment conditions. For instance, interaction between factors A and B is the contrast between the average of combined treatment (A + B) and no treatment on the one hand, and the average of the conditions A-only and B-only on the other hand. The two main effects are likewise contrasts between pairs of conditions. So in terms of analysis and sample size, each contrast in a twoway design is equivalent to that in a trial with two groups, with the other two contrasts as orthogonal covariates. The sample sizes for the fully cluster randomized 2 × 2 design and the fully person randomized 2 × 2 design can thus be computed with the Equations 1–8. The same holds for the main effects in a design with a cluster-level factor A and a person-level factor B, for which the optimal design and sample size can be computed as discussed in the sections Optimal Design of Cluster Randomized Trials and Optimal Design of Multisite/Multicenter Trials, respectively. The cross-level interaction of A and B requires some explanation. It can be tested by first computing per cluster the outcome difference between both levels of B, and then using that difference as outcome in an unpaired t-test comparing the two levels of A. Since this involves two unpaired samples of $K/2$ clusters each, the variance of this interaction is four times as large as of the main effect of B.

In short, all analyses of twoway nested designs with a quantitative outcome come down to applying model (1) or (7) and so the optimal design can be found just as for those models. For dichotomous outcomes, see Moerbeek and Maas (2005) who derive optimal twoway nested designs under the constraint of balancedness, that is, an equal number of clusters and of persons per cluster in each cell. Now, balanced allocation is not in general optimal for the logistic model, because of heterogeneity of outcome variance between treatment conditions, see Equation 11. As explained in the section Dichotomous Outcomes, the optimal allocation depends on the unknown fixed effects. However, at least for twoway non-nested designs with a dichotomous outcome, it was recently shown that balanced allocation is the best way to cope with this local optimality problem (Yang, Mandal, & Majumdar, 2012). Finally, optimal design for the simultaneous testing of all three contrasts in the twoway design is possible by minimizing the trace or the determinant of the covariance matrix of the three estimators. As explained in the Example section, this is known as A-optimality and D-optimality, respectively.

Nested Experiments With Two or Three Repeated Measures

In nested trials, just like in any trial, the dependent variable is often repeatedly measured, for instance before and after treatment, or after treatment and at a follow-up. Here, we focus on designs with two or three repeated measures.

Consider first the case of two repeated measures, a posttest and a follow-up. The treatment effect can be evaluated at either time point, but this makes sense only if there is treatment by time interaction. In the absence of such interaction, testing the treatment effect averaged across time points is more powerful and parsimonious. This leads to the following strategy for analysis and design, assuming absence of missing data: First, test treatment by time interaction by computing per person the difference DY between both measures and applying model (1) or (7) to DY. Secondly, if no interaction is found, then test the main effect of treatment by computing per person the average AY of both repeated measures and applying model (1) or (7) to AY. Finally, if interaction is found, test the treatment effect per time point by applying model (1) or (7) to each measure apart. Since all tests are based on model (1) or (7), the optimal design and sample size calculations in this paper also apply to the nested design with two repeated measures, noting that outcome variance, ICC, and effect size must now be specified in terms of DY and AY. Optimal design for the simultaneous estimation and testing of AY and DY is also possible by minimizing the trace (*A*-optimality) or the determinant (*D*-optimality) of the covariance matrix of the corresponding two effect estimators.

If the design also includes a pretest of the dependent variable, it can be included into the analysis either as a covariate or as a repeated measure. The first method is more powerful than the second (Rausch et al., 2003; Van Breukelen, 2006) and its effects on sample size and power have been discussed in the section Covariates. However, missing data call for an adapted analysis to prevent persons from dropping out of the analysis because of a missing measure, and this has consequences for the analysis and optimal design method. Using a three-level mixed regression, with repeated measures nested within persons within clusters or sites, allows inclusion of every person with at least one observation after treatment onset. Using the pretest as a repeated measure instead of as a covariate even allows inclusion of persons with both other measures missing. If there are no missing data, this method has somewhat lower power than the pretest-as-covariate approach, but this power loss can be recovered by a small model modification (Van Breukelen, 2006). Optimal design of three-level studies is addressed in the next section.

Nested Designs With Three Levels and Multiple Repeated Measures

In addition to two-level designs, three-level designs frequently occur. Two important cases are (a) nesting of pupils within classes within schools, and (b) nesting of repeated measures within persons within clusters. Optimal sample sizes for case (a), assuming random intercepts but fixed slopes, are given by Heo and Leon (2008), Moerbeek, Van Breukelen, and Berger (2000), and Teerenstra, Moerbeek, Van Achterberg, Pelzer, and Borm (2008). Their equations resemble those for two-level designs. In particular, the budget B affects the optimal sample size at the level of randomization, due to absence of a random slope. If the slope for the treatment effect varies randomly, then the budget B affects the optimal sample size at the highest level at which the slope varies (Moerbeek et al., 2008, p. 183).

Optimal design and sample sizes for case (b) are covered in Heo, Kim, Xue, and Kim (2010) and Heo and Leon (2009). Their model is a simple variance components model with fixed slopes and no autocorrelation and their work is technically close to that in Moerbeek et al. (2008) and in Teerenstra et al. (2008) for case (a). The design of cluster randomized trials with repeated measures is also considered by Feldman and McKinlay (1994) who take a mixed ANOVA approach, allowing the time effects to vary randomly between clusters and persons. These authors also consider the choice between a cohort and cross-sectional design, where a new sample of persons is drawn from the same included clusters at each time point.

Finally, studies of optimal design of non-nested experiments with repeated measures have some interesting implications for nested experiments. Winkens, Schouten, Van Breukelen, and Berger (2005, 2006, 2007) considered the optimal number and timing of repeated measures in a between-subject experiment with a fixed total follow-up time per person, where the optimality criterion was the power and precision for the treatment effect, that is, the outcome difference between treated and control group. The group difference was assumed to be linearly divergent (i.e., linearly increasing over time) or quadratically divergent (i.e., could increase, or first increase and then level off, or first increase and then decrease over time). Further, the correlation between repeated measures was either assumed to be the same for all pairs of measures (compound symmetry) or to decrease as the time interval increases (autoregressive error). Winkens et al. also considered the person-to-measurement cost ratio. Their results can be summarized as follows: Under compound symmetry, about five repeated measures is highly efficient or even optimal, with the first measurement at the start and the last at the end of the follow-up time. The optimal allocation of the intermediate time points is not equidistant, but at least for quadratic divergence equidistance is highly efficient. Under autoregressive correlation, we need two measures for linear divergence, and three equidistant measures for quadratic divergence, always with the first measure at the start and the last at the end of the follow-up time. If compound symmetry and autoregression are equally likely, then three measures is a better choice than five, except if the person-to-measure cost ratio is very high or if substantive dropout is anticipated.

Related results have been obtained by various authors for experimental and observational studies (e.g., Berger & Tan, 2004; Galbraith & Marshner, 2002; Hedeker, Gibbons,

& Waternaux, 1999; Lu, Mehrotra, & Liu, 2009; Maxwell, 1998; Ortega-Azurduy, Tan, & Berger, 2008, 2009; Ouwens, Tan, & Berger, 2002; Overall & Doyle, 1994; Raudenbush & Liu, 2001; Schouten, 1999; Tan & Berger, 1999; Tekle, Tan, & Berger, 2008a, 2008b; Tu, Kowalski, Zhang, Lynch, & Crits-Christoph, 2004; Venter, Maxwell, & Bolig, 2002; Yi & Panzarella, 2002).

Discussion

This paper introduced optimal design of experiments with nesting of persons within organizations, or nesting of repeated measures within persons. There are many other instances of multilevel design optimization, for instance in fMRI (nesting of repeated measures within brain regions within subjects) and in psychological testing (nesting of responses to successive test items within testees). See the corresponding papers elsewhere in this issue for more detail. There are also other methods of optimization, of which those aiming at a robust design will now be mentioned within the context of randomized experiments. As Equation 5 already showed, optimal design for estimating regression parameters in mixed linear models depends on unknown variance parameters. Optimal design for mixed logistic models for dichotomous data even depends on the unknown regression parameters themselves. This dependence is known as the local optimality problem in optimal design literature, since a design which is optimal for certain parameter values may be far from optimal for other values, and so misspecification of some parameters may lead to inefficient designs. There is quite some literature on what may be called "robust design," that is, on designs which suffer less from local optimality. Broadly speaking, there are three different approaches: Maximin design, Bayesian design, and sequential and adaptive design.

Maximin design maximizes either the minimum efficiency, or the minimum relative efficiency of the design relative to the optimal design, over a range of plausible parameter values, by choosing the design with the highest minimum. Some examples of this approach are Van Breukelen and Candel (2012) for cluster randomized and multisite trials, and Ouwens et al. (2002) and Tekle et al. (2008b) for longitudinal studies. Bayesian design maximizes not the minimum efficiency, but the expected efficiency, by integrating over a prior distribution for the unknown parameter(s) upon which the efficiency depends. Examples are Rotondi and Donner (2009), Spiegelhalter (2001), and Turner, Prevost, and Thompson (2004). Finally, group sequential design uses an initial sample size based upon a best guess of the parameter values and splits that sample into two or more groups or batches. Data analysis after each batch leads to a decision to continue the sampling, or to stop because of statistical significance of the effect of interest, or to stop because a significant effect is not be expected upon completion (futility stopping). Special test statistics and significance levels are used to control the type I and II error rates. Adaptive design of clinical trials differs from group sequential design in that the initial sample size is smaller, and that data analysis leads to a reestimation of the total sample size needed and may also lead to other design modifications. Chow and Chang (2007) give a good introduction into sequential and adaptive design of clinical trials. Lake, Kammann, Klar, and Betensky (2002), and Zou, Donner, and Klar (2005) discuss some methods in the context of cluster randomized trials.

As a final remark, common to all optimal design research is the aim to improve the cost-effectiveness of empirical studies, which is nice in a time of budget cuts. Since application of optimal design of multilevel studies is easier with software, here are some packages. MLPowSim (http://www.bristol.ac.uk/cmm/software/mlpowsim) creates R scripts and MLwiN macros for sample size calculation for various models with R or MlwiN. Similarly, ML-DEs (http://ppw.kuleuven.be/english/research/mesrg) provide R scripts to run simulation studies in MLwiN to compare the efficiencies of different multilevel designs under budget and cost constraints. Finally, the site http://sitemaker.mich.edu/group-based/optimal_design_software provides software by Raudenbush et al. to plot power against effect size, ICC, and sample size, for cluster randomized and multisite trials, and for longitudinal studies using specific models.

References

Ankenman, B. E., Aviles, A. I., & Pinheiro, J. C. (2003). Optimal designs for mixed-effects models with two random nested factors. *Statistica Sinica, 13*, 385–401.

Atkinson, A. C., Donev, A. N., & Tobias, R. D. (2007). *Optimum experimental designs, with SAS*. Oxford, UK: Oxford University Press.

Ausems, M., Mesters, I., Van Breukelen, G. J. P., & De Vries, H. (2002). Short-term effects of a randomized computer-based out-of-school smoking prevention trial aimed at Dutch elementary schoolchildren. *Preventive Medicine, 34*, 581–589.

Berger, M. P. F. (1986). A comparison of efficiencies of longitudinal, mixed longitudinal, and cross-sectional designs. *Journal of Educational Statistics, 11*, 171–181.

Berger, M. P. F., & Tan, F. E. S. (2004). Robust designs for linear mixed effects models. *Journal of the Royal Statistical Society C: Applied Statistics, 53*, 569–581.

Berger, M. P. F., & Wong, W. K. (2009). *An introduction to optimal designs for social and biomedical research*. Chichester, UK: Wiley.

Borm, G. F., Melis, R. J. F., Teerenstra, S., & Peer, P. (2005). Pseudo cluster randomization: A treatment allocation method to minimize contamination and selection bias. *Statistics in Medicine, 24*, 3535–3547.

Candel, M. J. J. M., & Van Breukelen, G. J. P. (2009). Varying cluster sizes in trials with clusters in one treatment arm: Sample size adjustments when testing treatment effects with linear mixed models. *Statistics in Medicine, 28*, 2307–2324.

Candel, M. J. J. M., & Van Breukelen, G. J. P. (2010). Sample size adjustments for varying cluster sizes in cluster randomized trials with binary outcomes analyzed with second-order PQL mixed logistic regression. *Statistics in Medicine, 29*, 1488–1501.

Chow, S. C., & Chang, M. (2007). *Adaptive design methods in clinical trials*. Boca Raton, FL: Chapman and Hall/CRC.

Cohen, M. P. (1998). Determining sample sizes for surveys with data analyzed by hierarchical linear models. *Journal of Official Statistics, 14*, 267–275.

DeCoster, J., Iselin, A. R., & Gallucci, M. (2009). A conceptual and empirical examination of justifications for dichotomization. *Psychological Methods, 14*, 349–366. doi: 10.1037/a0016956

Donner, A., Birkett, N., & Buck, C. (1981). Randomization by cluster: Sample size requirements and analysis. *American Journal of Epidemiology, 114*, 906–914.

Donner, A., & Klar, N. (2000). *Design and analysis of cluster randomization trials in health research*. New York, NY: Wiley.

Eldridge, S. M., & Kerry, S. (2012). *A practical guide to cluster randomised trials in health services research*. Chichester, UK: Wiley.

Fedorov, V., Mannino, F., & Zhang, R. (2009). Consequences of dichotomization. *Pharmaceutical Statistics, 8*, 50–61. doi: 10.1002/pst.331

Feldman, H. A., & McKinlay, S. M. (1994). Cohort versus cross-sectional design in large field trials: Precision, sample size and a unifying model. *Statistics in Medicine, 13*, 61–78.

Fox, J. (1997). *Applied regression analysis, linear models, and related methods*. Thousand Oaks, CA: Sage.

Galbraith, S., & Marshner, I. C. (2002). Guidelines for the design of clinical trials with longitudinal outcomes. *Controlled Clinical Trials, 23*, 257–273.

Giradeau, B., Ravaud, P., & Donner, A. (2008). Sample size calculation for cluster-randomized cross-over trials. *Statistics in Medicine, 27*, 5578–5585 (with correction in volume *28*, page 720).

Goos, P., & Jones, B. (2011). *Optimal design of experiments: A case study approach*. Chichester, UK: Wiley.

Hayes, R. J., & Moulton, L. H. (2009). *Cluster randomized trials*. Boca Raton, FL: Chapman and Hall.

Hedeker, D., Gibbons, R. D., & Waternaux, C. (1999). Sample size estimation for longitudinal designs with attrition: Comparing time-related contrasts between two groups. *Journal of Educational and Behavioral Statistics, 24*, 70–93.

Heo, M., Kim, Y., Xue, X., & Kim, M. Y. (2010). Sample size requirement to detect an intervention effect at the end of follow-up in a longitudinal cluster randomized trial. *Statistics in Medicine, 29*, 382–390.

Heo, M., & Leon, A. C. (2008). Statistical power and sample size requirements for three level hierarchical cluster randomized trials. *Biometrics, 64*, 1256–1262.

Heo, M., & Leon, A. C. (2009). Sample size requirements to detect an intervention by time interaction in longitudinal cluster randomized trials. *Statistics in Medicine, 28*, 1017–1027.

Hernandez, A. V., Steyerberg, E. W., & Habbema, J. D. F. (2004). Covariate adjustment in randomized controlled trials with dichotomous outcomes increases statistical power and reduces sample size requirements. *Journal of Clinical Epidemiology, 57*, 454–460.

Hsieh, F. Y., Bloch, D. A., & Larsen, M. D. (1998). A simple method of sample size calculation for linear and logistic regression. *Statistics in Medicine, 17*, 1623–1634.

Hussey, M. A., & Hughes, J. P. (2007). Design and analysis of stepped wedge cluster randomized trials. *Contemporary Clinical Trials, 28*, 182–191.

Imai, K., King, G., & Nall, C. (2009). The essential role of pair matching in cluster-randomized experiments, with application to the Mexican universal health insurance evaluation. *Statistical Science, 24*, 29–53.

Klar, N., & Donner, A. (1997). The merits of matching in community intervention trials: A cautionary tale. *Statistics in Medicine, 16*, 1753–1764. (plus letter and author reply in volume *17*, 2149–2152).

Kraag, G., Van Breukelen, G. J. P., Kok, G. J., & Hosman, C. (2009). "Learn Young, Learn Fair", a stress management program for fifth and sixth graders: Longitudinal results from an experimental study. *Journal of Child Psychology and Psychiatry, 50*, 1185–1195.

Lake, S., Kammann, E., Klar, N., & Betensky, R. (2002). Sample size re-estimation in cluster randomized trials. *Statistics in Medicine, 21*, 1337–1350.

Lu, K., Mehrotra, D. V., & Liu, G. (2009). Sample size determination for constrained longitudinal data analysis. *Statistics in Medicine, 28*, 679–699.

Maas, C. J. M., & Hox, J. J. (2005). Sufficient sample sizes for multilevel modeling. *Methodology, 1*, 86–92.

MacCallum, R. C., Zhang, S., Preacher, K. J., & Rucker, D. D. (2002). On the practice of dichotomization of quantitative variables. *Psychological Methods, 7*, 19–40. doi: 10.1037//1082-989X.7.1.19

Maxwell, S. E. (1998). Longitudinal designs in randomized group comparisons: When will intermediate observations increase statistical power? *Psychological Methods, 3*, 275–290.

McClelland, G. H. (1997). Optimal design in psychological research. *Psychological Methods, 2*, 3–19.

Moerbeek, M. (2005). Randomization of clusters versus randomization of persons within clusters: Which is preferable? *American Statistician, 59*, 72–78.

Moerbeek, M. (2006). Power and money in cluster randomized trials: When is it worth measuring a covariate? *Statistics in Medicine, 25*, 2607–2617.

Moerbeek, M., & Maas, C. J. M. (2005). Optimal experimental designs for multilevel logistic models with two binary predictors. *Communications in Statistics A – Theory and Methods, 34*, 1151–1167.

Moerbeek, M., Van Breukelen, G. J. P., & Berger, M. P. F. (2000). Design issues for experiments in multilevel populations. *Journal of Educational and Behavioral Statistics, 25*, 271–284.

Moerbeek, M., Van Breukelen, G. J. P., & Berger, M. P. F. (2001a). Optimal experimental designs for multilevel models with covariates. *Communications in Statistics – Theory and Methods, 30*, 2683–2697.

Moerbeek, M., Van Breukelen, G. J. P., & Berger, M. P. F. (2001b). Optimal experimental design for multilevel logistic models. *Journal of the Royal Statistical Society D: The Statistician, 50*, 17–30.

Moerbeek, M., Van Breukelen, G. J. P., & Berger, M. P. F. (2003). A comparison between traditional methods and multilevel regression for the analysis of multicentre intervention studies. *Journal of Clinical Epidemiology, 56*, 341–350.

Moerbeek, M., Van Breukelen, G. J. P., & Berger, M. P. F. (2008). Optimal designs for multilevel studies. In J. DeLeeuw & E. Meijer (Eds.), *Handbook of Multilevel Analysis* (pp. 177–205). New York, NY: Springer.

Moerbeek, M., & Wong, W. K. (2008). Sample size formula for trials comparing group and individual treatments in a multilevel model. *Statistics in Medicine, 27*, 2850–2864.

Moore, K. L., Neugebauer, R., Valappil, T., & Van der Laan, M. J. (2011). Robust extraction of covariate information to improve estimation efficiency in randomized trials. *Statistics in Medicine, 30*, 2389–2408.

Murray, D. M. (1998). *Design and analysis of group-randomized trials*. New York, NY: Oxford University Press.

Ortega-Azurduy, S. A., Tan, F. E. S., & Berger, M. P. F. (2008). The effect of drop-out on the efficiency of D-optimal designs of linear mixed models. *Statistics in Medicine, 27*, 2601–2617.

Ortega-Azurduy, S. A., Tan, F. E. S., & Berger, M. P. F. (2009). Highly efficient designs to handle the incorrect specification of linear mixed models. *Communications in Statistics – Simulation and Computation, 38*, 14–30.

Ouwens, J. N. M., Tan, F. E. S., & Berger, M. P. F. (2002). Maximin D-optimal design for longitudinal mixed effects models. *Biometrics, 58*, 735–741.

Overall, J. E., & Doyle, S. R. (1994). Estimating sample sizes for repeated measurement designs. *Controlled Clinical Trials, 15*, 100–123.

Peeters, F., Huibers, M., Roelofs, J., Van Breukelen, G. J. P., Hollon, S. D., Markowitz, J. C., ... Arntz, A. (2013). The clinical effectiveness of evidence-based interventions for depression: A pragmatic trial in routine practice. *Journal of Affective Disorders, 145*, 349–355.

Raudenbush, S. W. (1997). Statistical analysis and optimal design for cluster randomized trials. *Psychological Methods, 2*, 173–185.

Raudenbush, S. W., & Liu, X. (2000). Statistical power and optimal design for multisite trials. *Psychological Methods, 5*, 199–213.

Raudenbush, S. W., & Liu, X. (2001). Effects of study duration, frequency of observation, and sample size on power in studies of group differences in polynomial change. *Psychological Methods, 6*, 387–401.

Rausch, J. R., Maxwell, S. E., & Kelley, K. (2003). Analytic methods for questions pertaining to a randomized pretest, posttest, follow-up design. *Journal of Clinical Child and Adolescent Psychology, 32*, 467–486.

Rhoads, C. H. (2011). The implications of "contamination" for experimental design in education. *Journal of Educational and Behavioral Statistics, 36*, 76–104.

Rietbergen, C., & Moerbeek, M. (2011). The design of cluster randomized crossover trials. *Journal of Educational and Behavioral Statistics, 36*, 472–490.

Roberts, C., & Roberts, S. A. (2005). Design and analysis of clinical trials with clustering effects due to treatment. *Clinical Trials, 2*, 152–162.

Robinson, L. D., & Jewell, N. P. (1991). Some surprising results about covariate adjustment in logistic regression models. *International Statistical Review, 58*, 227–240.

Rotondi, M. A., & Donner, A. (2009). Sample size estimation in cluster randomized educational trials: An empirical Bayes approach. *Journal of Educational and Behavioral Statistics, 34*, 229–237.

Schouten, H. J. A. (1999). Planning group sizes in clinical trials with a continuous outcome and repeated measures. *Statistics in Medicine, 18*, 255–264.

Searle, S. (2006). *Linear models for unbalanced data* (2nd ed.). New York, NY: Wiley.

Searle, S., Casella, G., & McCulloch, C. (1992). *Variance components*. New York, NY: Wiley.

Searle, S., & Pukelsheim, F. (1986). Effect of intraclass correlation on weighted averages. *The American Statistician, 40*, 103–105.

Senn, S. J. (1989). The use of baselines in clinical trials of bronchodilators. *Statistics in Medicine, 8*, 1339–1350.

Senn, S. J. (1998). Some controversies in planning and analyzing multi-centre trials. *Statistics in Medicine, 17*, 1753–1765.

Senn, S. J., & Julious, S. (2009). Measurement in clinical trials: A neglected issue for statisticians? *Statistics in Medicine, 28*, 3189–3209. doi: 10.1002/sim.3603

Snijders, T. A. B., & Bosker, R. J. (1993). Standard errors and sample sizes for two-level research. *Journal of Educational Statistics, 18*, 237–259.

Snijders, T. A. B., & Bosker, R. J. (2011). *Multilevel analysis: An introduction to basic and advanced multilevel modeling* (2nd ed.). London, UK: Sage.

Spiegelhalter, D. J. (2001). Bayesian methods for cluster randomized trials. *Statistics in Medicine, 20*, 435–452.

Tan, F. E. S., & Berger, M. P. F. (1999). Optimal allocation of time points for the random effects model. *Communications in Statistics B: Simulation and computation, 28*, 517–540.

Teerenstra, S., Eldridge, S., Graff, M., DeHoop, E., & Borm, G. F. (2012). A simple sample size formula for analysis of covariance in cluster randomized trials. *Statistics in Medicine, 31*, 2169–2178.

Teerenstra, S., Moerbeek, M., Melis, R. J. F., & Borm, G. F. (2007). A comparison of methods to analyse continuous data from pseudo cluster randomized trials. *Statistics in Medicine, 26*, 4100–4115.

Teerenstra, S., Moerbeek, M., Van Achterberg, T., Pelzer, B. J., & Borm, G. F. (2008). Sample size calculations for 3-level cluster randomized trials. *Clinical Trials, 5*, 486–495.

Tekle, F. B., Tan, F. E. S., & Berger, M. P. F. (2008a). D-optimal cohort designs for linear mixed effects models. *Statistics in Medicine, 27*, 2586–2600.

Tekle, F. B., Tan, F. E. S., & Berger, M. P. F. (2008b). Maximin D-optimal designs for binary longitudinal responses. *Computational Statistics & Data Analysis, 52*, 5253–5262.

Tu, X. M., Kowalski, J., Zhang, J., Lynch, K. G., & Crits-Christoph, P. (2004). Power analyses for longitudinal trials and other clustered designs. *Statistics in Medicine, 23*, 2799–2815.

Turner, R. M., Prevost, A. T., & Thompson, S. G. (2004). Allowing for imprecision of the intracluster correlation coefficient in the design of cluster randomized trials. *Statistics in Medicine, 23*, 1195–1214.

Turner, R. M., White, I. R., & Croudace, T. (2007). Analysis of cluster randomized cross-over trial data: A comparison of methods. *Statistics in Medicine, 26*, 274–289.

Van Breukelen, G. J. P. (2006). ANCOVA versus change from baseline: More power in randomized studies, more bias in nonrandomized studies. *Journal of Clinical Epidemiology, 59*, 920–925.

Van Breukelen, G. J. P., & Candel, M. J. J. M. (2011). Efficient design of cluster randomized and multicentre trials with unknown intraclass correlation. *Statistical Methods in Medical Research*. Advance online publication. doi: 10.1177/0962280211421344

Van Breukelen, G. J. P., & Candel, M. J. J. M. (2012). Calculating sample sizes for cluster randomized trials: We can keep it simple and efficient! *Journal of Clinical Epidemiology, 65*, 1212–1218.

Van Breukelen, G. J. P., Candel, M. J. J. M., & Berger, M. P. F. (2007). Relative efficiency of unequal versus equal cluster sizes in cluster randomized and multicentre trials. *Statistics in Medicine, 26*, 2589–2603.

Van Breukelen, G. J. P., Candel, M. J. J. M., & Berger, M. P. F. (2008). Relative efficiency of unequal cluster sizes for variance component estimation in cluster randomized and multicentre trials. *Statistical Methods in Medical Research, 17*, 439–458.

Van Keulen, H., Mesters, I., Ausems, M., Van Breukelen, G. J. P., Campbell, M., Resnicow, K., ... DeVries, H. (2011). Tailored print communication and telephone motivational interviewing are equally successful in improving multiple lifestyle behaviors in an RCT. *Annals of Behavioral Medicine, 41*, 104–118.

Van Keulen, H., Mesters, I., Brug, J., Ausems, M., Campbell, M., Resnicow, K., ... DeVries, H. (2008). Vitalum study design: RCT evaluating the efficacy of tailored print communication and telephone motivational interviewing on multiple health behaviors. *BMC Public Health, 8*, 216.

Vargha, A., Rudas, T., Delaney, H. D., & Maxwell, S. E. (1996). Dichotomization, partial correlation, and conditional independence. *Journal of Educational and Behavioral Statistics, 21*, 264–282.

Venter, A., Maxwell, S. E., & Bolig, E. (2002). Power in randomized group comparisons: The value of adding a single intermediate time point to a traditional pretest-posttest design. *Psychological Methods, 7*, 194–209.

Winkens, B., Schouten, H. J. A., Van Breukelen, G. J. P., & Berger, M. P. F. (2005). Optimal time-points in clinical trials with linearly divergent treatment effects. *Statistics in Medicine, 24*, 3743–3756.

Winkens, B., Schouten, H. J. A., Van Breukelen, G. J. P., & Berger, M. P. F. (2006). Optimal number of repeated measures and group sizes in clinical trials with linearly divergent treatment effects. *Contemporary Clinical Trials, 27*, 57–69.

Winkens, B., Schouten, H. J. A., Van Breukelen, G. J. P., & Berger, M. P. F. (2007). Optimal designs for clinical trials with second-order polynomial treatment effects. *Statistical Methods in Medical Research, 16*, 523–537.

Yang, J., Mandal, A., & Majumdar, D. (2012). Optimal designs for two-level factorial experiments with binary response. *Statistica Sinica, 22*, 885–907.

Yi, Q., & Panzarella, T. (2002). Estimating sample size for tests on trends across repeated measurements with missing data based on the interaction term in a mixed model. *Controlled Clinical Trials, 23*, 481–496.

Zou, G. Y., Donner, A., & Klar, N. (2005). Group sequential methods for cluster randomization trials with binary outcomes. *Clinical Trials, 2*, 479–487.

Gerard J. P. van Breukelen

Faculty of Psychology and Neuroscience
Maastricht University
P.O. Box 616
6200 MD Maastricht
The Netherlands
Tel. +31 43 388-4001
E-mail gerard.vbreukelen@maastrichtuniversity.nl

Appendix A

Covariance Matrix and Optimal Design for a Cluster Randomized Trial

In the absence of covariates, ML estimation of the fixed effects reduces to weighted least squares (WLS) estimation using as observations the outcome means of the clusters (Searle, 2006). Under the joint assumptions of an equal sample size per cluster and homogeneity of variance, WLS in turn reduces to ordinary least squares (OLS) estimation (Searle & Pukelsheim, 1986). Assuming K clusters of which the first $K/2$ are assigned to control and the second $K/2$ to treatment, this gives the following simple estimator of the expected outcome under control and under treatment, respectively:

$$\hat{\mu}_c = \frac{2}{K}\sum_{j=1}^{K/2} \bar{Y}_j, \quad \hat{\mu}_t = \frac{2}{K}\sum_{j=1+K/2}^{K} \bar{Y}_j,$$

where $\bar{Y}_j$ is the observed mean outcome of all n individuals in cluster j.

It follows from Equation 1c that $\text{Var}\left(\bar{Y}_j\right) = \sigma_{u0}^2 + \frac{\sigma_e^2}{n}$. This in turn gives

$$\text{Var}(\hat{\mu}_c) = \text{Var}(\hat{\mu}_t) = \frac{n\sigma_{u0}^2 + \sigma_e^2}{nK/2} = \left(\frac{2\sigma_Y^2}{nK}\right)[(n-1)\rho + 1],$$

where $\sigma_y{}^2$ and ρ are the total outcome variance and the intraclass correlation in (2), respectively. Further, $\text{Cov}(\hat{\mu}_c, \hat{\mu}_t) = 0$ because $\hat{\mu}_c$ and $\hat{\mu}_t$ are obtained from independent samples. Finally, if treatment x_j is coded as -0.5 for control and $+0.5$ for treated ((0,1) coding is discussed later), we get the following fixed effects estimators:

$$\begin{pmatrix}\hat{\beta}_0\\ \hat{\beta}_1\end{pmatrix} = \begin{pmatrix}\hat{\mu}_t + \hat{\mu}_c\\ \hat{\mu}_t - \hat{\mu}_c\end{pmatrix}, \tag{A.1}$$

and so their covariance matrix equals (using the fact that $\text{Cov}(\hat{\mu}_c, \hat{\mu}_t) = 0$):

$$\text{Cov}\begin{pmatrix}\hat{\beta}_0\\ \hat{\beta}_1\end{pmatrix} = \begin{bmatrix}\text{Var}(\hat{\mu}_t) + \text{Var}(\hat{\mu}_c) & \text{Var}(\hat{\mu}_t) - \text{Var}(\hat{\mu}_c)\\ \text{Var}(\hat{\mu}_t) - \text{Var}(\hat{\mu}_c) & \text{Var}(\hat{\mu}_t) + \text{Var}(\hat{\mu}_c)\end{bmatrix},$$

which, since $\text{Var}(\hat{\mu}_c) = \text{Var}(\hat{\mu}_t)$, reduces to

$$\text{Cov}\begin{pmatrix}\hat{\beta}_0\\ \hat{\beta}_1\end{pmatrix} = \begin{bmatrix}1 & 0\\ 0 & 1\end{bmatrix} \times \left(\frac{4\sigma_Y^2}{nK}\right)[(n-1)\rho + 1]. \tag{A.2}$$

Minimizing either the trace, or the determinant, of this matrix, or simply $\text{Var}\left(\hat{\beta}_1\right)$, means minimizing the term on the right side of the multiplication sign in (A.2).

If treatment x_j is coded as 0 for control and 1 treated, we get instead:

$$\begin{pmatrix}\hat{\beta}_0\\ \hat{\beta}_1\end{pmatrix} = \begin{pmatrix}\hat{\mu}_c\\ \hat{\mu}_t - \hat{\mu}_c\end{pmatrix},$$

which results in:

$$\text{Cov}\begin{pmatrix}\hat{\beta}_0\\ \hat{\beta}_1\end{pmatrix} = \begin{bmatrix}\text{Var}(\hat{\mu}_c) & -\text{Var}(\hat{\mu}_c)\\ -\text{Var}(\hat{\mu}_c) & \text{Var}(\hat{\mu}_t) - \text{Var}(\hat{\mu}_c)\end{bmatrix},$$

which can be written into:

$$\mathrm{Cov}\begin{pmatrix}\hat{\beta}_0\\ \hat{\beta}_1\end{pmatrix} = \begin{bmatrix} 0.5 & -0.5 \\ -0.5 & 1 \end{bmatrix} \times \left(\frac{4\sigma_Y^2}{nK}\right)[(n-1)\rho+1]. \tag{A.3}$$

Again, minimizing the trace, or the determinant, or Var $(\hat{\beta}_1)$, means minimizing the term on the right side of the multiplication sign. So let us consider minimization of Var $(\hat{\beta}_1)$.

Filling in the cost constraint $B = K(c + sn)$ gives us:

$$\mathrm{Var}(\hat{\beta}_1) = \left(\frac{4\sigma_Y^2}{nB}\right)[(n-1)\rho+1](c+sn). \tag{A.4}$$

Setting the partial derivative of (A.4) with respect to n equal to zero gives as solution the sample size n in (5), which can be verified to give the minimum of Var $(\hat{\beta}_1)$. The corresponding number of clusters K follows from the budget constraint in (4).

Appendix B

Derivation of the Sample Size Equation for a Cluster Randomized Trial

Consider at first the classical non-nested randomized trial (between-subject experiment) for testing $H0$: $\mu_1 = \mu_2$ against $H1$: $\mu_1 \neq \mu_2$ for a quantitative outcome with known variance σ^2 (the case of unknown variance gives the same derivation in terms of a t-test instead of a z-test). Under H0 the following test statistic has a standard normal distribution, assuming a sample size n in each group and total sample size $N = 2n$:

$$Z_0 = \frac{(\bar{Y}_1 - \bar{Y}_2)}{SE}, \text{ where } SE = \sqrt{\frac{2\sigma^2}{n}} = \sqrt{\frac{4\sigma^2}{N}}.$$

Denoting as $z_{1-\alpha/2}$ the critical value of the test statistic, and assuming that, in truth $\mu_1 < \mu_2$, the test power $(1-\gamma)$ satisfies

$$(1-\gamma) = P(Z_0 < z_{1-\alpha/2}),$$

which equals

$$P\left(Z_1 < z_{1-\alpha/2} - \frac{(\mu_1-\mu_2)}{SE}\right),$$

where $Z_1 = \frac{(\bar{Y}_1 - \bar{Y}_2) - (\mu_1 - \mu_2)}{SE}$ has a standard normal distribution under $H1$. From this it follows that

$$z_{1-\gamma} = z_{1-\alpha/2} - \frac{(\mu_1-\mu_2)}{SE},$$

which can be rewritten into the total sample size equation

$$N = 4(z_{1-\gamma} + z_{1-\alpha/2})^2 \times \left(\frac{1}{d}\right)^2. \tag{B.1}$$

This result can be translated into Equation 6 for cluster randomized trials in two ways.

First, one may apply (B.1) to cluster means, with N being the number of clusters K, and d being the effect size in terms of cluster means. Using next the equation $\mathrm{Var}(\bar{Y}_j) = \sigma_{u0}^2 + \frac{\sigma_e^2}{n} = (\frac{\sigma_Y^2}{n})((n-1)\rho+1)$, it can be verified that the squared effect size d^2 for cluster means is DE/n times as large as $d^2 = (\frac{\mu_1-\mu_2}{\sigma_Y})^2$ for individual data, where DE = $(n-1)\rho + 1$ is the design effect. Rewriting Equation B.1 in terms of the d^2 for individual data then gives Equation 6 for the number of clusters needed.

Another derivation is this: Let N in (B.1) be the total number of individuals, nK, needed if the ICC is zero. Then multiply this total number with the design effect DE, and finally divide both sides of the equation by the sample size per cluster, n.

Appendix C

Covariance Matrix and Optimal Design for a Multisite Trial

For the multisite trial model of Equation 7d the derivations are as follows:First, within site j estimate the expected outcome per treatment group, given assignment of the first $n/2$ individuals to control and of the second $n/2$ individuals to treatment:

$$\hat{\mu}_{cj} = \frac{2}{n}\sum_{i=1}^{n/2} Y_{ij}, \quad \hat{\mu}_{tj} = \frac{2}{n}\sum_{i=1+n/2}^{n} Y_{ij}.$$

Next, assuming that treatment is coded −0.5 for control and +0.5 for treatment, we estimate the intercept and slope per site as follows:

$$\begin{pmatrix}\hat{\beta}_{0j}\\ \hat{\beta}_{1j}\end{pmatrix} = \begin{pmatrix}\hat{\mu}_{tj} + \hat{\mu}_{cj}\\ \hat{\mu}_{tj} - \hat{\mu}_{cj}\end{pmatrix},$$

which are, respectively, the sum and the difference between treated and controls within site j. Finally, the fixed effect estimators are obtained by averaging the site-specific estimators over the K sites, giving:

$$\begin{pmatrix}\hat{\beta}_0\\ \hat{\beta}_1\end{pmatrix} = \frac{1}{K}\sum_{j=1}^{K}\begin{pmatrix}\hat{\mu}_{tj} + \hat{\mu}_{cj}\\ \hat{\mu}_{tj} - \hat{\mu}_{cj}\end{pmatrix}, \tag{C.1}$$

and

$$\mathrm{Cov}\begin{pmatrix}\hat{\beta}_0\\ \hat{\beta}_1\end{pmatrix} = \frac{4}{nK}\begin{bmatrix} n\sigma_{u0}^2 + \sigma_e^2 & 0 \\ 0 & n\frac{1}{4}\sigma_{u1}^2 + \sigma_e^2 \end{bmatrix}, \tag{C.2}$$

under the assumption of no slope-intercept covariance (σ_{u0u1} = 0), noting that such covariance would imply heterogeneity of variance between treatment groups, which is beyond the present scope. Defining now the following two 'intraclass correlations',

$$\rho_0 = \frac{\sigma_{u0}^2 + \sigma_e^2}{\sigma_y^2} \quad \text{and} \quad \rho_1 = \frac{\frac{1}{4}\sigma_{u1}^2 + \sigma_e^2}{\sigma_y^2},$$

we can rewrite the covariance matrix of fixed effect estimators into:

$$\mathrm{Cov}\begin{pmatrix}\hat{\beta}_0 \\ \hat{\beta}_1\end{pmatrix} = \frac{4\sigma_y^2}{nK}\begin{bmatrix}(n-1)\,\rho_0 + 1 & 0 \\ 0 & (n-1)\,\rho_1 + 1\end{bmatrix},$$

which is almost the same as for the cluster randomized trial, and shows how the optimal design and sample size equations for the cluster randomized trial can be applied to the multisite trial. In particular, if $\mathrm{Var}\,(\hat{\beta}_1)$ is the optimality criterion, we need to use $\rho 1$ instead of the ICC definition of Equation 2.

Just as for the cluster randomized trial, we can derive the covariance matrix for the multisite trial with (0, 1) coding of treatment. However, this coding scheme gives heterogeneity of variance unless $\sigma_{u0u1} = \frac{-\sigma_1^2}{2}$, and so the results are more complicated.

Original Article

Optimal Designs for Event History Analysis

Mirjam Moerbeek and Katarzyna Jóźwiak

Department of Methodology and Statistics, Utrecht University, The Netherlands

Abstract. The aim of event history analysis is to study the occurrence and timing of events, such as premature psychotherapy termination or drinking onset, and to relate the probability of event occurrence to relevant covariates. In the social sciences, the timing of the event is often measured discretely by using time intervals, which implies the exact timing of event occurrence is unknown. The optimal number of subjects and time intervals need to be decided upon in the design stage of a trial with discrete-time survival endpoints. This paper shows how the optimal design depends on the underlying survival function and the costs to include a subject relative to the costs to take a measurement. Furthermore, the effects of attrition on the optimal design are studied. An example on drinking onset illustrates the proposed methodology.

Keywords: longitudinal study, optimal design, study duration, sample size, time intervals

An important research question in psychological research is if and when subjects experience certain events. An example is the relapse to drinking of recently treated alcoholics. In such a study one could relate the occurrence and timing of relapse to potentially important covariates such as age, gender, alcohol abuse history, and socioeconomic status, with the aim to identify those that are at highest risk of relapse. Also, one could identify the points in time with highest risk of relapse and monitor patients more closely during these time periods. Another example is the termination of psychotherapy by psychiatric patients. As premature termination may result in a decline of mental health, it is important to identify those patients who are at highest risk of treatment termination and to develop interventions to adhere them to therapy for a sufficient amount of time. A third example is a study on the relationship between social status and the timing of aggressive behavior, for instance of subjects waiting for a traffic light.

Studies on the occurrence and timing of events are very common in the biomedical sciences and survival from some disease is the main event that is studied. For this reason, the statistical analysis technique is generally referred to as survival analysis. In the social and psychological sciences, on the other hand, a variety of events can be studied, such as entry into parenthood, drop-out from college, or premature termination of psychotherapy. In these fields of science, less ill-tempered terminology, such as event history analysis, is often used.

Survival analysis has various approaches. First, one can study single versus repetitive events. A single event is an event that can happen only once, such as the birth of someone's first child. A repetitive event can occur multiple times in one's life, such as starting smoking after having quit smoking previously. In this case, the interest lies not only in the timing one starts smoking for the very first time but also in starting smoking after some period of abstinence. Second, a distinction can be made between survival analysis models that do account for competing events and those that do not. In many cases of survival analysis one is interested in the timing of one single event but there are competing events that preclude the observation of the event of interest. An example is the study on the onset of some disease where for some subjects the onset cannot be observed because they died from another disease prior to the onset. Third, the precision with which the timing of the event can be measured is important. In the biomedical sciences the timing of an event can often be measured continuously using thin and precise units, such as days. In the social sciences the event time is more often measured discretely by using coarse time intervals, such as weeks, months, or even years.

It is obvious a loss of information occurs in the latter case and an enumeration of reasons why one would measure time discretely rather than continuously is given here. First, in retrospective studies the respondents cannot always remember the exact timing of the event very precisely. For instance, a smoker may be able to remember the year in which he or she started smoking but not the exact date. Second, for ethical, logistical, administrative, and financial reasons it is not always possible to contact respondents too frequently in prospective studies. In a smoking prevention intervention to prevent and delay the onset of smoking, for instance, one cannot contact participants on a daily basis and will only do so less frequently, say once each quarter. Third, some events can only occur at discrete points in time. Graduation from university, for example, can only

DOI: 10.1027/2151-2604/a000144

occur at a limited number of preset dates during the academic year.

The choice for the metric for time is an important one since it determines the statistical model to be used for data analysis. The Cox proportional hazard model (Cox, 1972) is the standard model for continuous-time survival data. Discrete-time survival data can be analyzed by means of logistic regression models. A logistic regression model relates the probability of event occurrence in one single time interval to predictor variables. With discrete-time survival models, a logistic regression model is formulated for each time interval, which are combined in a single equation model (Singer & Willett, 1993, 2003).

The metric of time also determines the design of a trial with event history data. The main design issue for continuous-time survival analysis is how many subjects should be included. This question has been extensively studied in the past couple of decades and sample size recommendations can be found in textbooks on survival analysis (Collett, 2003, Hosmer, Lemeshow, & May, 2008, Machin, Cheung, & Parmar, 2006). Sample sizes are calculated in such a way that a sufficient number of events may be expected to occur during the course of the trial. For discrete-time survival analysis, not only has the number of subjects to be decided upon but also the number of times they need to be measured. Increasing both quantities implies more precise estimates of the predictor variables' effects but also results in elongated study duration and higher costs of the study. It is therefore important to select the optimal combination of number of subjects and number of measurements in such a way that the available budget is not exceeded.

Optimal designs for discrete-time survival analysis have been studied only recently (Jóźwiak & Moerbeek, 2012). The aim of this paper is to give a nontechnical introduction to such optimal designs for trials with single, nonrecurring, and noncompeting events. Graphical representations show how the optimal design can be selected, how the performance of suboptimal designs can be evaluated, and how optimal designs depend on the underlying survival function and on the ratio of the costs to include a subject relative to the costs to take a measurement. A real-life example is used to illustrate the methodology of this paper. The contents of this paper are as follows. The next section introduces the example and poses questions that are important in the design phase of a trial with discrete-time survival endpoints. The following section gives a short rehearsal of logistic regression models for dichotomous outcome variables and it shows how multiple logistic regression models can be combined into one single model for discrete-time survival data. The optimal design methodology is presented in the next section and is followed by a section of results. An example on drinking onset in adolescents is used to illustrate the proposed methodology. Attention is also paid to a computer program that facilitates the calculation of optimal designs. The final section gives some concluding remarks and a request to all researchers who perform trials with discrete-time survival endpoints.

Illustrative Example: Drinking Onset

Early onset of drinking has been shown to be associated with problematic drinking in adulthood. For this reason, various alcohol prevention programs have been developed to prevent children and adolescents from starting drinking at an early age. The effects of such programs can be evaluated in a randomized controlled trial where part of the participants are randomized to a drinking prevention program (the intervention condition) and the others do not receive such a program (the control condition). The main outcome in such a trial is the timing of drinking onset and since it may be too expensive to contact all participants on a daily basis one may measure the outcome in discrete-time intervals, such as years.

If there exists an effect of the intervention one would like to detect it with sufficient probability. In other words, one would like to design the trial in such a way that it has sufficient power for the test on treatment effect. Power increases with increasing sample size, but also the number of measurements per subject has an effect on power. An important question is what is the optimal combination of sample size and number of subjects. Another important question is what power for the test on treatment effect can be achieved by a proposed design, as well as the costs of a design to achieve a desired power level.

In the next two sections our focus is on models for discrete-time survival analysis and optimal designs for a Weibull survival function. We will then return to the example on drinking onset and show the power levels that can be achieved with a fixed sample size, the required sample size to achieve 80% power and the related costs of the study. These values are given for various durations of the study and enable to study the optimal combination of sample size and study duration.

Event History Analysis Models

Logistic regression models relate a categorical outcome variable to relevant predictor variables with the aim to estimate the effects of these predictors and to test for their significance. Our focus is on logistic regression models for dichotomous outcomes Y, which can take only two values: Event occurrence ($Y = 1$) versus nonoccurrence ($Y = 0$). The probability of event occurrence is denoted $h = P(Y = 1)$ and called the hazard probability. Instead of predicting the value of the outcome variable Y, the probability h of Y occurring is predicted by means of a logistic regression model. First we focus on the model that measures the outcome variable at one single occasion in time. For subject i

$$\text{logit}(h_i) = \log\left(\frac{h_i}{1 - h_i}\right) = \alpha + \beta_1 x_{1i} + \ldots + \beta_k x_{ki}, \quad (1)$$

where $x_{1i}, \ldots, x_{ki}$ are the predictor variables and $\beta_1, \ldots, \beta_k$ are their effects on the logit scale. The intercept α

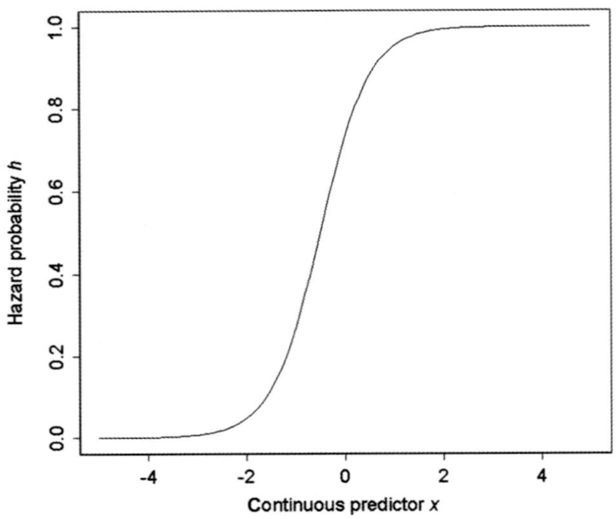

Figure 1. Example logistic regression curve.

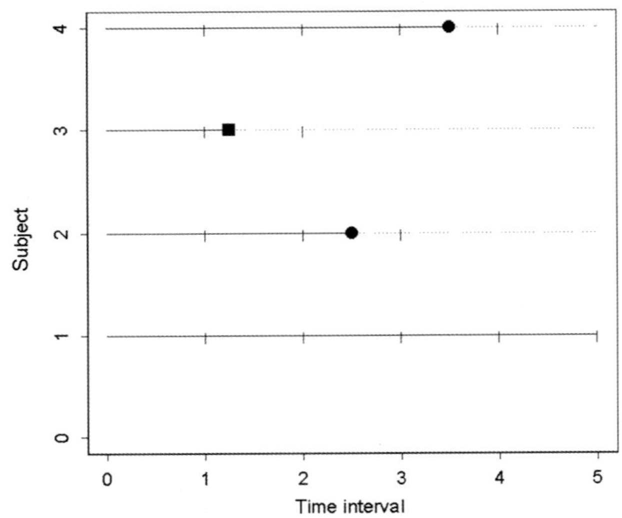

Figure 2. Graphical representation of follow-up times for four subjects in a trial with five time intervals.

determines the logit of the hazard probability if all predictors are equal to zero, and is called the baseline hazard probability. The logit is the natural logarithm of the probability of event occurrence divided by the probability of event nonoccurrence.

The regression model 1 shows a nonlinear relation between hazard probability and predictors. In fact, this relation is S-shaped and a subject's predicted hazard probability lies between 0 and 1 (as it should do by definition). A value close to 1 means that it is very likely the event has occurred, while a value close to 0 means it is very unlikely it has occurred. Figure 1 shows a hypothetical relation between the hazard probability h_i and a single continuous predictor x_i. The intercept α and predictor effect β values are equal to 1 and 2, respectively. A larger value for the intercept implies a larger hazard probability for any value of the predictor x_i. A larger value for the predictor effect implies a stronger relation between the predictor and outcome and hence a steeper slope of the curve in Figure 1. A negative value implies a declining hazard probability h_i as x_i increases.

For a more extensive introduction to logistic regression models, including evaluations of goodness of fit and tests on single or multiple regression weights, the reader is referred to Chapter 8 of Field (2009) or Chapter 10 of Tabachnick and Fidell (2013).

With discrete-time survival analysis each subject is followed for multiple time intervals, until and including the interval he or she experiences the event or gets censored because the study concludes or because of attrition due to unforeseen reasons other than event occurrence. In therapeutic trials, for instance, patients may get lost to follow up because they change therapist and lose contact with their former therapist or the study coordinator. The validity of a survival analysis rests on the assumption that censoring is noninformative, which implies that a subject's probability of getting censored does not depend on the event status. In that case, the censored subjects do not differ systematically from the noncensored ones, and it can therefore be assumed that all subjects who remain in the study after censoring are representative of everyone who would have remained in the study had there not been censoring. Since a statistical analysis cannot produce unbiased analyses in case of informative censoring, the advice is to work hard to prevent problems arising from informative censoring before they occur (Singer & Willett, 2003), for instance by giving the subjects incentives each time they report their event status at the end of a time interval. We hope that future advances in (multiple) imputation techniques will provide tools for dealing with noninformative censoring in a justified way.

Figure 2 gives a graphical representation of follow-up in a study with five time intervals. This figure depicts the time intervals on the horizontal axis and four subjects on the vertical. The solid horizontal lines represent the amount of time until either event occurrence (depicted as •) or attrition (depicted as ■). The dotted horizontal line represents the additional time the subject could have stayed in the study had he or she not experienced the event or dropped out from the study due to attrition. The tiny vertical solid lines represent observation times. Subject 1 remains in the study until the end of the study duration and has not experienced the event by then neither dropped out. Subject 2 experiences the event in time interval 3. The occurrence of the event is measured at time point 3 and the subject is not followed any further afterwards. Subject 3 drops out because of attrition in time interval 2, which is measured at time point 2. No further measurements are taken beyond this time point. Subject 4 experiences the event in time interval 4, as is measured at time point 4 but not at further time points.

Discrete-time survival data can be analyzed by means of a logistic regression model if they are presented in a so-called person-period dataset. For each subject this dataset contains rows of data for each time interval until and including the time interval the event or censoring occurs.

The person-period dataset for the small example in Figure 2 is given in Figure 3. It includes five rows for subject 1, three for subject 2, two for subject 3, and four for subject 4. The event status takes the value 1 for event occurrence and 0 for event nonoccurrence or attrition. Treatment condition, with 1 denoting the intervention and 0 denoting the control, is an example of a predictor variable and, if necessary, additional predictors may by included by expanding the number of columns. The dummy variables in the last five columns represent the time intervals and take on the value 1 in the time interval they represent and the value 0 otherwise.

Equation 2 is the discrete-time hazard model for subject i in time interval j in a study with J intervals and k predictors:

$$\text{logit}(h_{ij}) = \log\left(\frac{h_{ij}}{1-h_{ij}}\right) = \alpha_1 D_{1ij} + \ldots + \alpha_J D_{Jij} + \beta_1 x_{1ij} + \ldots + \beta_k x_{kij}. \quad (2)$$

This model is an extension of model 1 as it includes multiple intercepts $\alpha_1, \ldots, \alpha_J$, one for each of the time intervals. The collection of these intercepts is called the logit baseline hazard probability function. It should be noted that this model is called a proportional odds model since the effects of the predictors do not vary across time intervals (the β coefficients do not have a subscript j). A more extensive model would allow these effects to vary across time intervals, but is not discussed in this paper.

Standard statistical software for binary logistic regression can be applied to estimate the model parameters along with their standard errors and to test for their significance, provided the data are presented in the person-period format. An extensive introduction can be found in Singer and Willett (1993, 2003).

Optimal Design Methodology

Now that the basics of discrete-time survival analysis and the related statistical model have been introduced, it is time to focus on optimal designs. The focus of this paper is on optimal designs for randomized controlled trials, where subjects are randomly assigned to either a control or experimental condition. Random assignment and large sample sizes ensure that the treatment groups are comparable at baseline with respect to background variables that may have an effect of event occurrence and timing. In that case a difference between hazard probabilities across the treatment groups may be assigned to treatment rather

ID	Interval	Event status	Treatment	D1	D2	D3	D4	D5
1	1	0	1	1	0	0	0	0
1	2	0	1	0	1	0	0	0
1	3	0	1	0	0	1	0	0
1	4	0	1	0	0	0	1	0
1	5	0	1	0	0	0	0	1
2	1	0	0	1	0	0	0	0
2	2	0	0	0	1	0	0	0
2	3	1	0	0	0	1	0	0
3	1	0	1	1	0	0	0	0
3	2	0	1	0	1	0	0	0
4	1	0	0	1	0	0	0	0
4	2	0	0	0	1	0	0	0
4	3	0	0	0	0	1	0	0
4	4	1	0	0	0	0	1	0

Figure 3. Person-period dataset for the trial in Figure 2.

than other sources or variation. Furthermore, covariates are not taken into account and the model becomes

$$\text{logit}(h_{ij}) = \log\left(\frac{h_{ij}}{1-h_{ij}}\right) = \alpha_1 D_{1ij} + \ldots + \alpha_J D_{Jij} + \beta x_i. \quad (3)$$

Treatment condition is denoted x_i and takes the value 0 for the control condition and the value 1 for the experimental. As it does not depend on the time interval indicator j, treatment condition does not change over time and hence the effects of crossover cannot be studied. Such a predictor is called a time-invariant predictor.

A study should be designed in such a way it provides most information on the research question(s), given the collected data and the chosen regression model to relate the outcome variable to relevant predictors. A design is determined by various design factors which each have a number of levels. For trials with discrete-time survival data the design is determined by the number of subjects, the number of measurements per subject, and the proportion of subjects who are allocated to the experimental condition. The first two design factors can take positive integer values; the third can take any value between 0 and 1. The amount of information that is provided by a design is captured by the Fisher information matrix and its inverse is the covariance matrix of the vector of model parameters $\theta = (\alpha_1, \ldots, \alpha_J, \beta)$.' This covariance matrix contains the variances and covariances of the parameters' estimators. The optimal design is the combination for which the Fisher information is maximized, or, in other words, for which the covariance matrix is minimized. Since matrices cannot be ordered in a unique way, various optimality criteria have been proposed, such as *D*-optimality, which minimizes the determinant of the covariance matrix. The optimality criterion should reflect the research question of the trial at hand. In most trials the primary research question is the difference between experimental conditions with respect to the outcome variable and the $\text{Var}(\hat{\beta})$, the variance of the treatment effect estimator $\hat{\beta}$, is used as optimality criterion. This is an obvious choice since minimal variance results in smallest confidence interval width and highest power of the test on treatment effect. The power of a test is the probability to detect a significant treatment effect, provided such an effect exists in the population. The power should be of sufficient size since not being able to detect an existing effect may be regarded as a waste of time and money of the researchers and their participants. Had another optimality criterion been used, the optimal design would not necessarily have resulted in highest power for the test on treatment effect. It should furthermore be mentioned that the optimality criterion $\text{Var}(\hat{\beta})$ does not focus on the efficient estimation of the baseline logit hazard probability function $(\alpha_1, \ldots, \alpha_J)$. If this function is also of interest, for instance because one wishes to estimate the probabilities of event occurrence in each time period in the control condition in an efficient way, then an optimality criterion that does involve the baseline hazard probability function should be taken into account.

The derivation of the optimal design involves matrix algebra and no simple mathematical expression to derive the optimal design can be given. For this reason the optimal designs are presented graphically. A mathematical justification for the derivation of the optimal design can be found in the appendix and is recommended to the interested reader with a sufficient knowledge of matrix algebra. From this appendix it follows that the optimal design depends on the model parameters $\alpha_1, \ldots, \alpha_J$, and β and such a design is called a locally optimal design.

One would expect a more efficient design if the sample size and/or the number of time intervals increase. In practice, these two quantities cannot increase without bounds for two reasons. First, a trial often has to be concluded within a certain amount of time. For instance, a maximum of six monthly measurements can be taken in a trial that has to be concluded within half a year, while at most 12 monthly measurements can be taken in a trial with duration of at most one year. So, increasing the number of measurements implies increased study duration. Second, costs are associated with including and treating subjects and with taking measurements and the total of these costs should not exceed the budget. Finding the optimal design implies finding the design that has smallest $\text{Var}(\hat{\beta})$ among all designs with a number of time intervals that does not exceed the maximum number of time intervals and costs that do not exceed the budget.

The costs of a trial are determined by the costs at the subject and the costs at the measurement level. The costs at the subject level are independent of the number of measurements taken and consist of recruitment costs, treatment costs, administrative costs, and so forth. The costs at the measurement level are assumed to be constant across time intervals and examples are travel costs of the patient to the therapist, the costs of a therapy session, and the costs to take measurements. Four methods to calculate the costs of a trial are distinguished. With the first cost function all subjects are followed during the complete course of the trial even though they may experience the event before the trial concludes. This cost function assumes attrition to be absent. An example is a school-based smoking prevention intervention where all pupils complete questionnaires at all follow-up occasions even though they may have experienced the primary event (smoking initiation) before the end of the trial. In such a trial the administrative costs of only selecting those pupils who are still at risk could outweigh the costs of processing questionnaires of all pupils in the trial. With the second cost function the aim is also to follow all subjects until the end of the trial but some leave due to attrition. With the third cost function, subjects leave the study once they experience the event and no further costs are associated with taking measurements after event occurrence. An obvious example is a trial that studies survival from some disease and no further measurements are taken once the patient has died. With the fourth cost function subjects do not only leave the study due to event occurrence but also due to attrition. As with the third method, no further measurements are taken after leaving the study.

To be able to find the optimal design the survival function in the control condition needs to be known to calculate

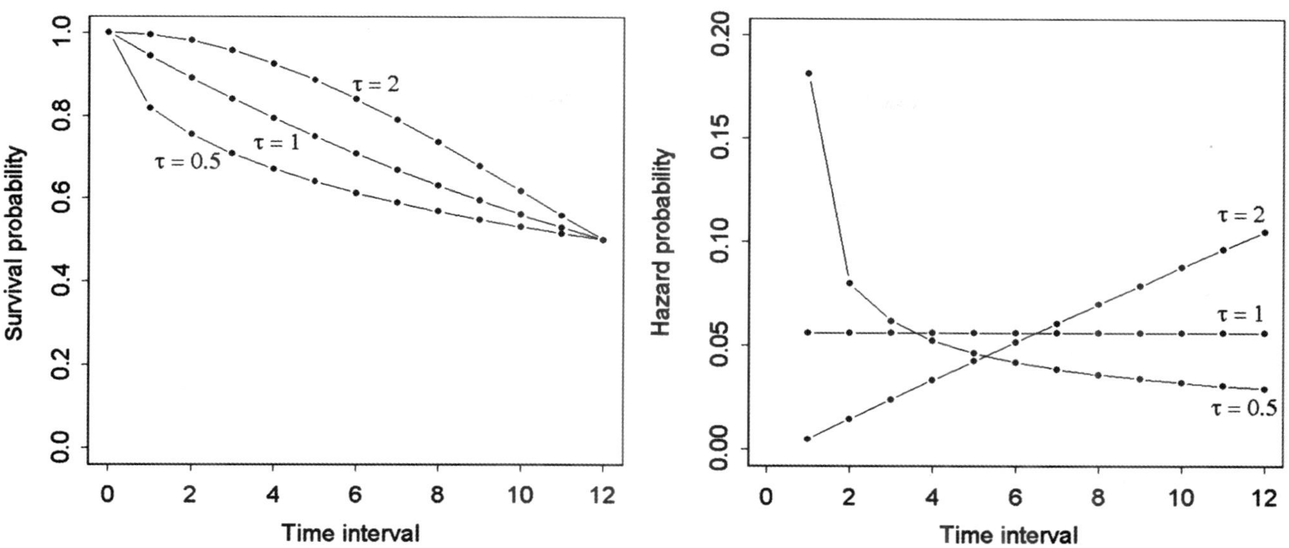

Figure 4. Example Weibull survival functions and related hazard probability functions in discrete time.

the baseline hazard probability function. The number of survival functions is infinite in theory and in the next section the focus is on the Weibull survival function, but our optimal design methodology applies to any survival function. The Weibull survival function can model constant, increasing or decreasing hazard probabilities across time and is characterized by two parameters: ω and τ. The first represents the proportion of subjects who have experienced the event by the end of a trial. The second is a shape parameter: A value $\tau > 1$ implies the risk of event occurrence is concentrated toward the end of the trial while a value $\tau < 1$ means highest hazard probabilities are found at the beginning of the trial. A value $\tau = 1$ implies a constant hazard probability and this survival process is generally known as exponential survival. It should be mentioned that the underlying Weibull survival function is related to the logistic regression function in Equation 3. For an increasing hazard probability across time ($\tau > 1$) the intercepts $\alpha_1, \ldots, \alpha_J$ are an increasing function, for a constant hazard probability ($\tau = 1$) these intercepts are constant, and for a decreasing hazard probability ($\tau < 1$) these intercepts are also decreasing.

Examples of Weibull survival functions are given in the left panel of Figure 4 for a trial with 50% event occurrence at the end of time interval 12 and various values of the shape parameter. A survival function gives the proportion of subjects who have not yet experienced the event as a function of time. By definition, the value at baseline is equal to 1 and the survival function cannot increase. The underlying survival function is assumed continuous (as represented by the solid lines) but measurements are only taken at the end of each time interval (as represented by the closed circles). The related hazard probability functions are given in the right panel of Figure 4 and these either increase, decrease of remain constant across time.

The efficiency of a design relative to the optimal design is calculated as the ratio of their $\mathrm{Var}(\hat{\beta})$ values. This ratio varies between 0 and 1 and the value 1 corresponds to the optimal design. The inverse relative efficiency determines how often a design should be replicated to do as well as the optimal design in terms of $\mathrm{Var}(\hat{\beta})$. For instance, a relative efficiency equal to 0.5 implies the design should be replicated twice. It is obvious relative efficiency values close to 0.8 or even 0.9 are preferred.

Results

For ease of presentation it is assumed both treatment groups are of equal size at baseline, although of course any allocation ratio may be used in a randomized controlled trial. The interested reader is referred to Jóźwiak and Moerbeek (2013c) for a further discussion on the efficiency of unequal allocation ratios. In this section we study the optimal combination of number of measurements and number of subjects. In case the number of measurements is fixed beforehand, a power analysis suffices to calculate the required number of subjects. It should be mentioned that each time interval has the same width which is chosen at the beginning of the trial. The maximum number of time periods that is feasible for the trial at hand should be chosen and the maximum number of time periods and time interval width determine the duration of the study. The proportion ω of subjects who have experienced the event by the end of the study, of course, depends on study duration.

Cost Function 1

Figure 5 shows the relative efficiency of designs with up to 12 time intervals for the first cost function. The relative efficiency is shown as a function of the number of time intervals but not as a function of the number of subjects.

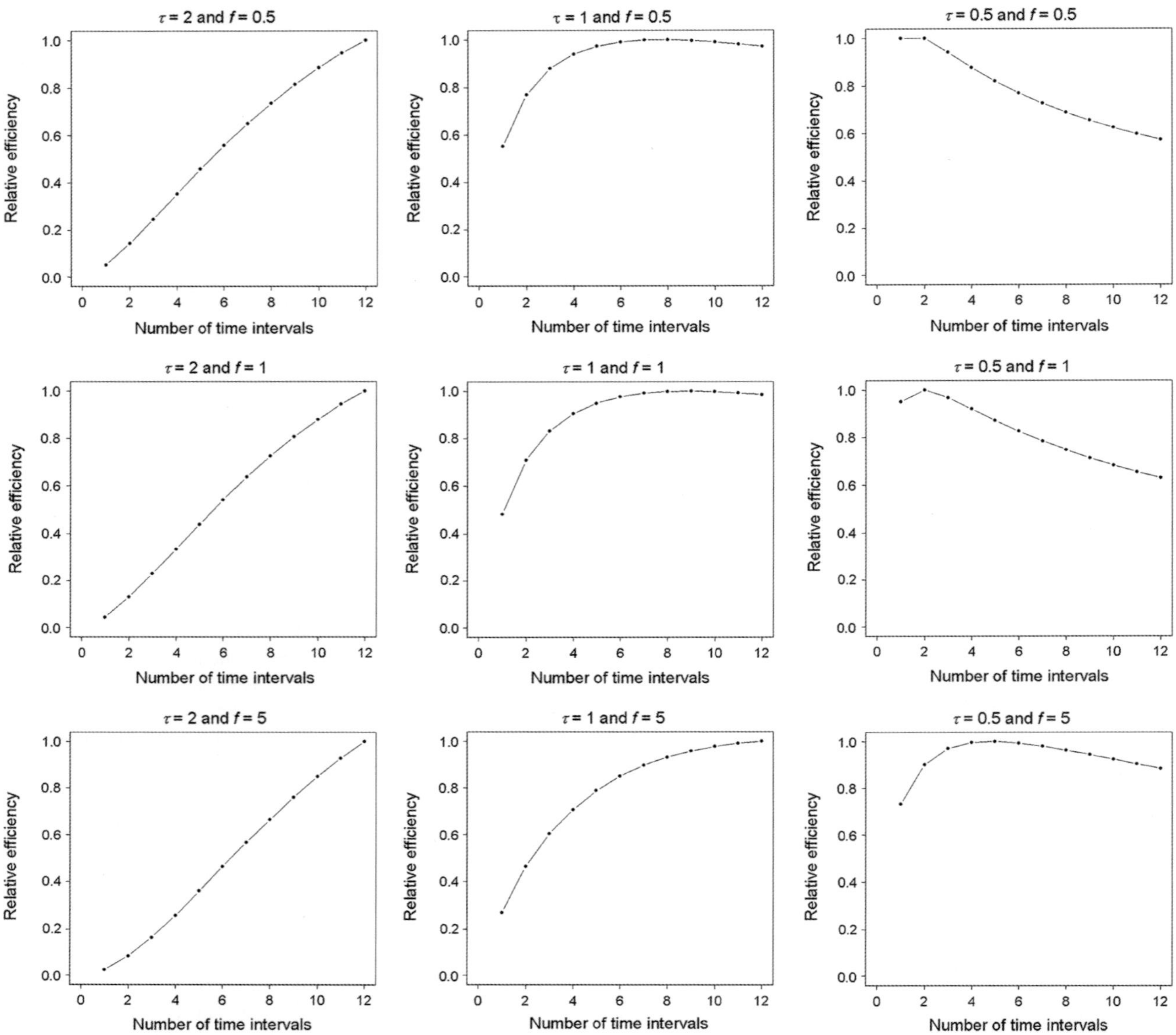

Figure 5. Efficiency plots for trials with a Weibull survival function and a maximum of 12 time intervals as based on the first cost function.

Once the optimal number of time intervals is determined, the optimal number of subjects follows from the available budget. The budget, in its turn, does not influence the relation between the number of time intervals and the relative efficiency. Multiplying the budget by some factor implies the number of subjects is multiplied by the same factor. Furthermore, it can be shown that the optimal design depends on the ratio of the costs at the subject level to the costs at the measurement level, rather than the costs themselves. This ratio is henceforward denoted f. For instance, if recruitment to the study and administrative costs are €250 per subject and each measurement costs €50 then this ratio is $f = 250/50 = 0.5$. The survival function also has an effect on the relative efficiency, but mainly through the shape parameter τ. The effects of the proportion event occurrence ω and the treatment effect β on the relative efficiency are very small and are not further studied. For a further justification of these findings, see the appendix and Jóźwiak and Moerbeek (2012).

Figure 5 is based on values $\omega = 0.5$ and $\beta = -0.5$, which imply 50% event occurrence by the end of time interval 12 and a smaller probability of event occurrence in the intervention condition than in the control. The columns of Figure 5 represent three different values τ. From left to right, these represent increasing, constant, and decreasing hazard probabilities. The rows of Figure 5 represent three different values of the cost ratio f. From top to bottom, these represent lower, equal, and higher costs per subject than per measurement.

The left column of Figure 5 shows that for $\tau = 2$, the optimal design has 12 time intervals. In other words, a maximum amount of time intervals is needed when the probability of event occurrence becomes larger toward the end of the trial. Longer study duration ensures more events occur during the course of the trial. For $\tau = 1$ or $\tau = 0.5$ the optimal number of time intervals either remains constant or increases with the cost ratio. In other words, more measurements, and hence fewer subjects, need to be taken in many cases where the costs at the subject level increase relative to the costs at the measurement level. Figure 5 also shows that for any given cost ratio f, the optimal number of time intervals decreases when the shape parameter τ decreases. In other words, fewer time intervals are needed when the probability of event occurrence is concentrated toward the beginning of the study duration. This effect is weakest for the largest cost ratio f.

The relative efficiencies of designs that are not optimal also follow from Figure 5 and strongly depend on the shape parameter and cost ratio. For a trial with event occurrence concentrated toward the end of the study duration (i.e., $\tau = 2$) the design efficiency strongly declines when the number of time intervals decreases from 12 to lower values and this effect is found for all three cost ratios. A less dramatic decline of the design efficiency can be observed in other scenarios, for instance the one with a shape parameter equal to $\tau = 0.5$ and a cost ratio $f = 5$, where any design with a number of time intervals larger than 1 has a reasonable efficiency larger than 0.8. In general it is observed that many of the alternative designs have a reasonable efficiency when the optimal design has a medium number of time intervals in the range from 1 to 12.

Table 1. Optimal number of time intervals as a function of the shape parameter and cost ratio

	$\tau = 2$	$\tau = 1$	$\tau = 0.5$
$f = 0.5$	12	8	2
	12	8	2
	12	12	2
	12	12	2
$f = 1$	12	9	2
	12	10	2
	12	12	3
	12	12	3
$f = 5$	12	12	5
	12	12	5
	12	12	8
	12	12	9

Note. The entries within each cell give the optimal number of time intervals for cost function 1, cost function 2, cost function 3, and cost function 4, respectively. Furthermore, $\omega = 0.5$ and $\beta = -0.5$.

The Effect of the Cost Function

Thus far, the focus was on the first cost function, which assumes all subjects are measured until the end of the study. The next step is to study the effects of various costs functions on the optimal number of time intervals. Keep in mind that with the second cost function all subjects are followed until attrition, with the third cost function all subjects are followed until event occurrence and that with the fourth cost function all subjects are followed until either event occurrence or attrition. As on average fewer measurements are taken per subject, more subjects can be recruited with the same budget.

Although the probability of attrition may vary across time intervals, a constant attrition rate throughout the course of the study is used and the probability of attrition between any two adjacent time points is fixed to be equal to 0.05. Also, noninformative attrition is assumed, which implies dropping out from the trial due to other reasons than event occurrence does not depend on event status.

Table 1 shows the optimal number of time intervals for the four cost functions and as a function of the shape parameter τ and cost ratio f. It is observed that termination of measurement after event occurrence or attrition may result in a higher optimal number of time intervals. For a trial with event occurrence concentrated toward the end of the study duration ($\tau = 2$) the optimal number of time intervals is equal to 12 for any cost function, so termination of follow-up after event occurrence or attrition does not influence the optimal design. For constant ($\tau = 1$) and increasing ($\tau = 0.5$) hazard probabilities termination of follow-up does influence the optimal design and in some cases the duration of the trial is almost doubled when the fourth cost function is used rather than the first.

Robustness of Optimal Designs

The true values of the model parameters need to be known to be able to calculate the optimal design. In other words, the optimal design is locally optimal. A researcher planning a trial with discrete-time survival endpoints should be able to specify the size of the treatment effect β and the baseline hazard probabilities. The baseline hazard probabilities follow from the underlying survival function in the control condition, which for the Weibull survival function is characterized by the parameters ω and τ. All model parameter values should be specified beforehand. This causes a vicious circle since the aim of a trial is to gain insight into the size of the unknown model parameters (in particular the size of the treatment effect) while the values of all model parameters need to be known in advance to be able to calculate the optimal design. One way to escape this circle is by replacing the treatment effect by the minimally relevant treatment effect and by replacing the baseline hazard probability function by a prior estimate. Such an estimate may be based on expert knowledge or findings from similar studies in the past. Of course, the prior estimate may be incorrect and it is worthwhile to study the robustness of

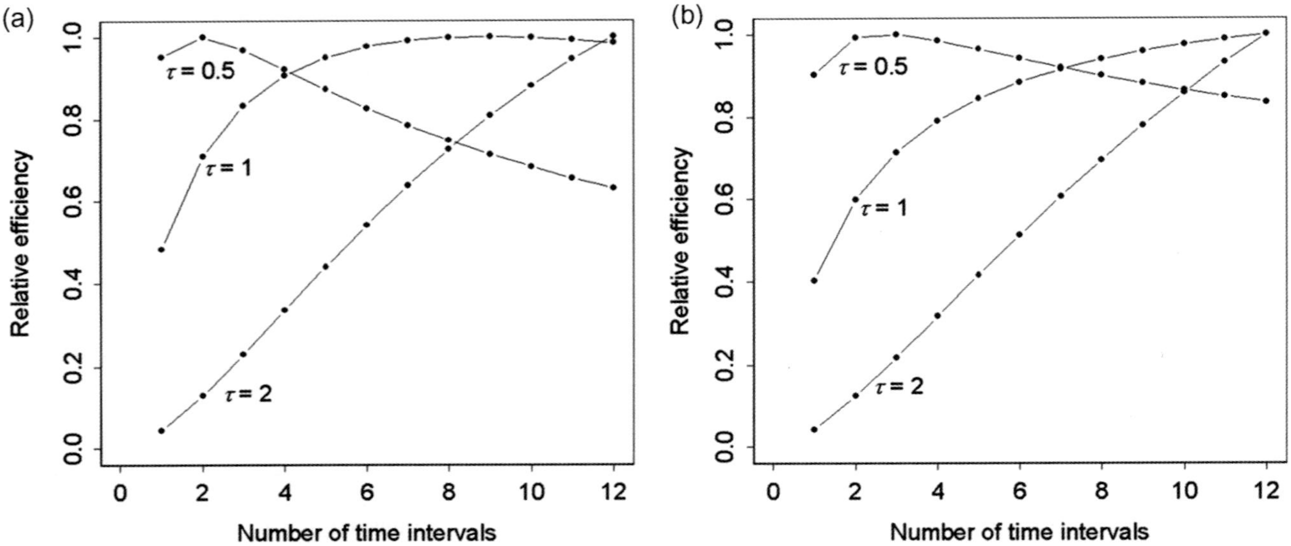

Figure 6. Relative efficiencies for three τ values and $f = 1$ (with $\omega = 0.5$ and $\beta = -0.5$) as based on the first cost function (a) and the third cost function with probability of attrition equal to 0.05 per time interval (b).

the optimal design against incorrect prior estimates of the model parameters.

Figure 6 presents relative efficiencies for designs with various values of the shape parameter to represent increasing ($\tau = 2$), constant ($\tau = 1$), and decreasing ($\tau = 0.5$) hazard probabilities across time. For both panels the following parameter values are used: $\omega = 0.5$ and $\beta = -0.5$. The left panel corresponds to the first cost function. A design with just two time intervals is optimal if one expects a decreasing hazard probability across time but the efficiency of this design is below 0.8 if the true hazard probability increases or remains constant across time. Vice versa, 12 time intervals are required when the hazard probability is expected to increase across time but this design is inefficient if the true hazard probability decreases across time. In summary, a design that performs reasonably well for all underlying hazard probability functions cannot be found for the first cost function. A more encouraging result is found for the third cost function, for which an efficiency of at least 0.8 can be achieved with at least ten time intervals for all three values of the shape parameter τ.

Optimal Design for the Illustrative Example

This section shows optimal designs and power levels for a trial on the effects of a drinking prevention program in adolescence and early adulthood. To be able to calculate the optimal design for a trial like this, the underlying survival probability function in the control condition needs to be known. For this purpose we use data from a study by Trim, Schuckit, and Smith (2010) that focused on the occurrence and timing of drinking onset of offspring from the San Diego prospective study. The study included 147 subjects whose drinking behavior was measured once each year from the age of 11 to the age of 22. This function and the related logit hazard probability function are shown in Figure 7. Furthermore, the expected size of the treatment effect β needs to be specified before the optimal design can be calculated. In general this effect size is unknown in the design phase of a trial and may be replaced by the minimally relevant effect size. Let us consider an effect size $\beta = -0.5$ to be relevant. This effect size implies 52% of the subjects in the intervention condition started drinking before the age of 22 years, which is a considerable decrease as compared to the 69% drinking onset in the control condition. The survival and logit hazard probability functions for the intervention condition are also presented in Figure 7 and a higher survival and lower hazard probability are observed in the intervention condition than in the control. The difference between the logit hazard probability functions remains constant across time as a proportion odds model is assumed. For both treatment groups, the logit hazard probability function zig-zags over time and the underlying survival function is clearly not Weibull. This does not pose a problem, however, since the optimal design methodology can be used for any underlying survival function, not just the Weibull survival function.

Table 2 gives results for a trial with a total of 294 subjects (i.e., 147 subjects per group, assuming both groups are of equal size at baseline). This table is based on the second cost function, that is, the aim is to measure all adolescents until the end of the trial but some leave earlier due to attrition. The attrition rate is assumed constant across time intervals and equal to 0.05. The power of a statistical test on treatment effect as a function of age at termination of the trial is given in the second column. As is obvious, the

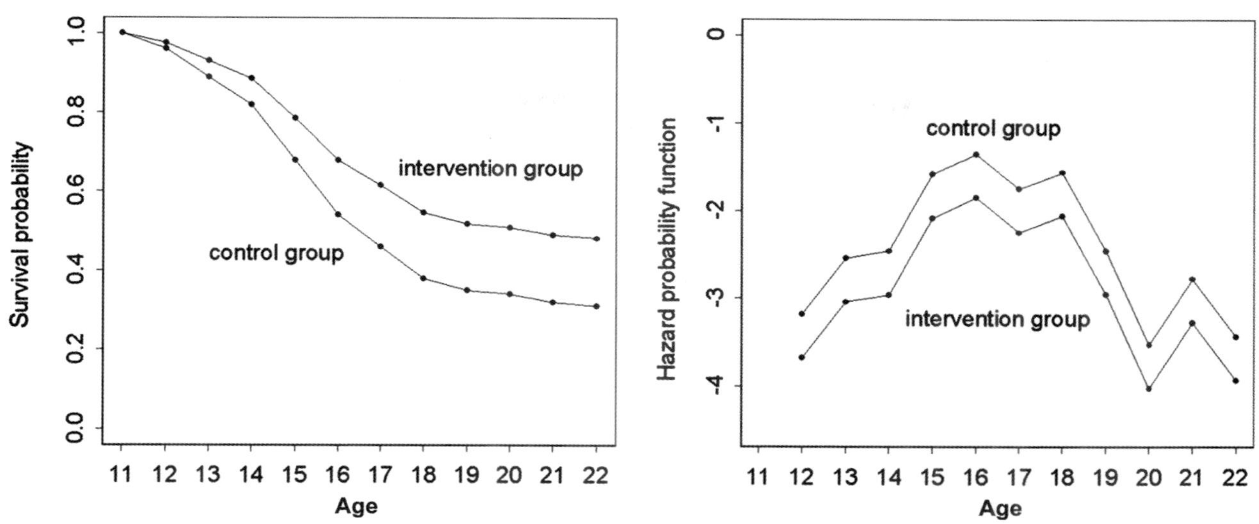

Figure 7. Survival and logit hazard probability functions for the example on drinking onset.

power increases when the trial ends at higher ages of the participants, and a sufficient power level of 80% is achieved when subjects are measured until the age of 22 years (i.e., for 11 time intervals). The third column shows how many subjects are needed to achieve 80% power for various values of age. Fewer subjects are needed when measurements at higher ages are included and a total sample size of 292 is required with a study that includes subjects up to the age of 22.

The next step is to calculate the costs of the study to achieve a power level 0.8. These costs are given in the last three columns of Table 2 and they are rescaled such that the costs to take a measurement are equal to 1. It is observed that for all costs ratios the costs are lowest when subjects are followed to the age of 18 years, that is, for seven time intervals. Taking a design that has slightly fewer or more time intervals than the optimal design has only a small effect on the trial costs. In any case, designs with just a few number of time intervals should be avoided since they result in very high costs.

Optimal Design Software

As was mentioned before, the optimal designs are calculated using matrix algebra and no simple closed-form formulae to calculate the optimal design can be given. For this reason, we developed the computer program PODSE (Program for Optimal Designs of Trials with Discrete-time Survival Endpoints) to facilitate researchers to calculate the optimal design for their trial with discrete-time survival endpoints. This is a stand-alone computer program with a graphical user interface and includes all four cost functions and a large variety of survival functions. It can handle three scenarios: Calculation of the optimal number of time

Table 2. Optimal designs for the drinking onset example

			Total costs of the study		
Age	Power with n = 294	n to get power = 0.8	f = 0.5	f = 1	f = 5
12	0.1107	4,248	20,815	12,532	5,905
13	0.2234	1,606	10,768	6,187	2,522
14	0.3289	1,004	8,453	4,729	1,749
15	0.5030	598	6,009	3,304	1,139
16	0.6406	430	4,986	2,708	886
17	0.7074	368	4,808	2,588	812
18	0.7618	324	4,686	2,505	760
19	0.7809	310	4,895	2,602	768
20	0.7870	304	5,183	2,744	792
21	0.7979	296	5,401	2,849	807
22	0.8031	292	5,661	2,976	829

Note. The column label age shows until which age the respondents are measured in the trial.

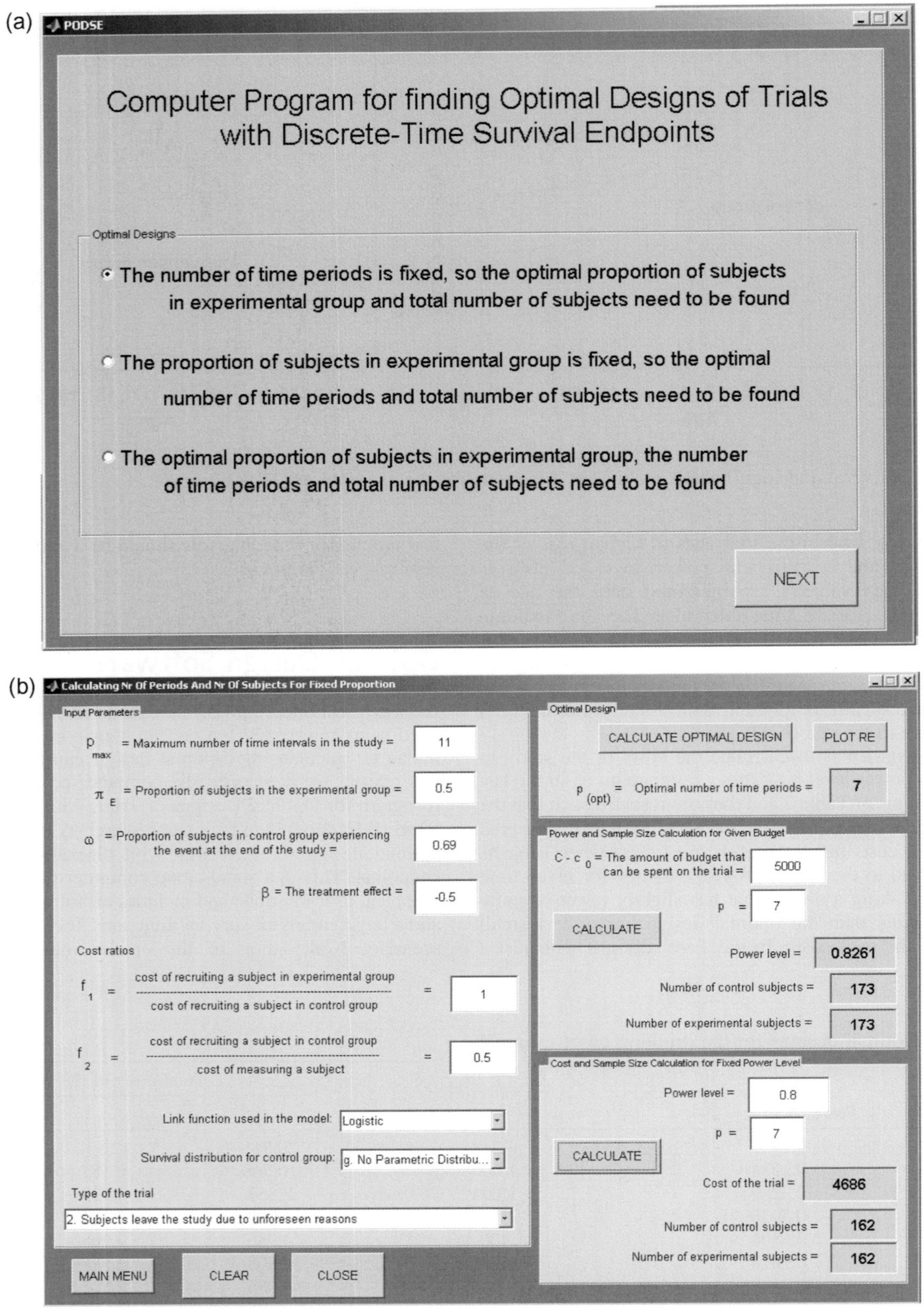

Figure 8. Menus of the PODSE program: main menu (a) and menu for trial with a fixed allocation to treatment conditions (b).

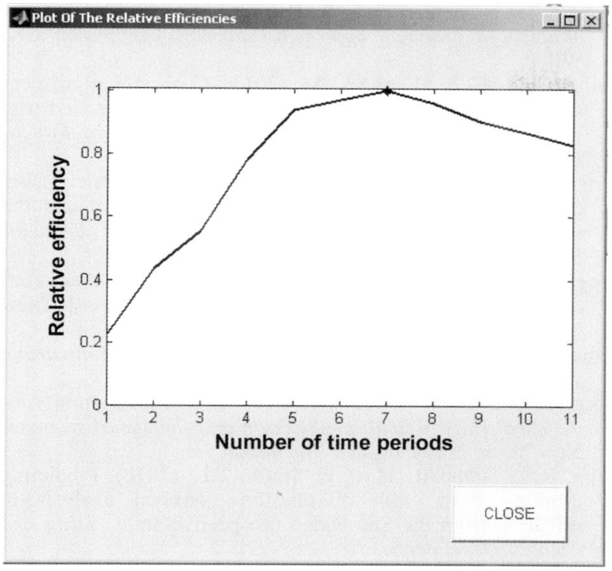

Figure 9. Relative efficiencies for alternative designs for the drinking onset example as calculated using the PODSE program.

intervals for a fixed allocation to treatment conditions, calculation of the optimal allocation when the number of time intervals is fixed beforehand, or calculation of the optimal design when neither the allocation nor the number of time intervals is fixed beforehand. Relative efficiency plots can be used to evaluate the performance of designs that are not optimal. For an extensive introduction of this program we refer to Jóźwiak and Moerbeek (2013b).

The PODSE program can be downloaded from http://tinyurl.com/discrete-survival. The main menu (Figure 8) that appears once the program has been started allows the user to select one of the three scenarios described above and after pressing the NEXT button a screen for the selected scenario appears. We discuss the use of PODSE for the scenario where the allocation ratio is fixed and the optimal number of time periods needs to be calculated. The bottom of Figure 8 shows the graphical user interface for this scenario. The left side allows the user to specify input needed to calculate the optimal design; the values that have been entered correspond to the example in the previous section. It should be mentioned that the cost function is an extension to the one presented in this paper as it allows the costs to take a measurement to vary across treatment conditions. Also, the program does not only allow for logistic regression but also for a model with a complementary log-log link function. The survival function in the control group can follow a parametric distribution (exponential, linear-exponential, Weibull, modified Weibull, log-logistic, or Gompertz) or a nonparametric distribution. In the latter case a small screen appears that allows the user to specify the baseline hazard function. The cost function can be selected at the bottom of the left half of the screen. In case of attrition (cost function 2 or 4) a small screen appears that allows the user to specify the attrition rate in each time period.

The optimal design is presented at the top of the right side. It is shown that the optimal design has $p = 7$ time periods, and this agrees with our finding in the previous section. A plot of the relative efficiencies of using alternative numbers of time periods at the same trial budget can be requested (Figure 9) and we observe that a high efficiency is achieved with a large range of alternative designs. The middle panel at the right allows the user to calculate the power level and number of that subjects that can be achieved with a fixed budget at a chosen number of time periods. Vice versa, the lower panel at the right is used for the calculation of the required budget to achieve a desired power level.

Concluding Remarks

The aim of this paper was to give an introduction to the methodology of optimal designs for trials with discrete-time survival endpoints and to show how the optimal design is determined by the shape of the underlying survival function and cost ratio. Of course, an introductory paper cannot cover all aspects of optimal designs. For instance, the possibility of unequal treatment group sizes at baseline was mainly ignored, while there are various reasons to use unequal group sizes (Dumville, Hahn, Miles, & Torgerson, 2006; Torgerson & Torgerson, 2008). Unequal group sizes may be preferred when the costs in the one condition are much larger than in the other, when attrition is much higher in the one condition than in the other, when the new treatment condition is likely to have harmful side effects, or when researchers expect subjects may benefit from the positive effects of the new treatment. Optimal allocation to treatment conditions can be found in Jóźwiak and Moerbeek (2013c). Another extension to our optimal design might be the inclusion of relevant covariates, which is of importance when the total sample size is small so randomization does not ensure the treatment groups do not differ with respect to relevant covariates. The effect of covariate adjustment on statistical power to detect an effect of treatment has been studied recently (Safarkhani & Moerbeek, 2013) and the effect of including covariates on optimal design is part of our current research. Finally, another focus is on trials were the time of recruitment varies across subjects while all subjects are followed until the same point in time at which the study concludes. This implies that those who enter later are followed for a shorter amount of time (Jóźwiak & Moerbeek, 2013a).

This paper concludes with a request to all researchers who perform trials with discrete-time survival endpoints. As prior estimates are needed to calculate the optimal design, it is important that estimates of the underlying survival probability or hazard probability function are clearly reported in the literature. As such, substantive researchers who are planning a trial with discrete-time survival endpoints are facilitated in their search for prior estimates of the baseline hazard probability function.

Acknowledgment

This study was funded by the Netherlands Organization for Scientific Research (NWO), Grant Number 451-08-004.

References

Collett, D. (2003). *Modelling survival data in medical research* (2nd ed.). New York, NY: Chapman and Hall/CRC.

Cox, D. (1972). Regression models and life tables. *Journal of the Royal Statistical Society Series B, 34*, 187–220.

Dumville, J. C., Hahn, S., Miles, J. N. V., & Torgerson, D. J. (2006). The use of unequal randomisation ratios in clinical trials: A review. *Contemporary Clinical Trials, 27*, 1–12.

Field, A. (2009). *Discovering statistics using SPSS*. London, UK: Sage.

Hosmer, D., Lemeshow, S., & May, S. (2008). *Applied survival analysis. Regression modeling of time-to-event data* (2nd ed.). Hoboken, NJ: Wiley.

Jóźwiak, K., & Moerbeek, M. (2012). Cost-effective designs for trials with discrete-time survival endpoints. *Computational Statistics and Data Analysis, 56*, 2086–2096.

Jóźwiak, K., & Moerbeek, M. (2013a). Accrual by cohorts in trials with discrete-time survival endpoints. *Clincal Trials, 10*, 33–42.

Jóźwiak, K., & Moerbeek, M. (2013b). PODSE: A computer program for optimal design of trials with discrete-time survival endpoints. *Computer Methods and Programs in Biomedicine, 111*, 115–127.

Jóźwiak, K., & Moerbeek, M. (2013c). Optimal treatment allocation and study duration for trials with discrete-time survival endpoints. *Journal of Statistical Planning and Inference, 143*, 971–982.

Machin, D., Cheung, Y., & Parmar, M. (2006). *Survival analysis: A practical approach* (2nd ed.). Chichester, UK: Wiley.

Safarkhani, M., & Moerbeek, M. (2013). Covariate adjustment strategy increases power in the randomized controlled trial with discrete-time survival endpoints. *Journal of Educational and Behavior Statistics, 38*, 355–380.

Singer, J. D., & Willett, J. B. (1993). It's about time: Using discrete-time survival analysis to study duration and the timing of events. *Journal of Educational and Behavior Statistics, 18*, 155–195.

Singer, J. D., & Willett, J. B. (2003). *Applied longitudinal data analysis. Modeling change and event occurrence*. New York, NY: Oxford University Press.

Tabachnick, B.G., & Fidell, L.S. (2013). *Using multivariate statistics*. Boston, MA: Pearson.

Torgerson, D. J., & Torgerson, C. J. (2008). *Designing randomised trials in health, education and the social sciences*. New York, NY: Palgrave Macmillan.

Trim, R.S., Schuckit, M.A., & Smith, T.L. (2010). Predicting drinking onset with discrete-time survival analysis in offspring from the San Diego prospective study. *Drug and Alcohol Dependence, 107*, 215–220.

Mirjam Moerbeek

Department of Methodology and Statistics
Utrecht University
Padualaan 14
3584CH Utrecht
The Netherlands
Tel. +31 30 253-1450
Fax +31 30 253-5797
E-mail M.Moerbeek@uu.nl

Appendix

Derivation of Optimal Designs

The aim of this appendix is to show how optimal designs are derived for trials with discrete-time survival endpoints. It is assumed the reader has a basic knowledge of matrix algebra. The statistical model for a trial with $j = 1, \ldots, J$ time periods and $i = 1, \ldots, n$ subjects is given by

$$\text{logit}(h_{ij}) = \log\left(\frac{h_{ij}}{1-h_{ij}}\right) = \alpha_1 D_{1ij} + \ldots + \alpha_J D_{Jij} + \beta x_i,$$

where α_j is the logit baseline hazard probability in time period j. Treatment condition x_i is constant within subjects and takes the value 0 for the control condition and the value 1 for the intervention condition. The effect of treatment on the logit scale is denoted β.

The model can be rewritten in matrix notation as

$$\text{logit}(h) = X\theta,$$

where h is a vector of probabilities for all observed subjects and time periods, and $\theta = (\alpha_1, \ldots, \alpha_J, \beta)'$ is a vector of unknown model parameters. X is the design matrix with $(J + 1)$ columns and $\sum_{j=1}^{J} n_{Cj} + \sum_{j=1}^{J} n_{Ej}$ rows where n_{Cj} and n_{Ej} are the number of subjects in the experimental and control condition entering time period j.

The estimated asymptotic covariance matrix of the maximum likelihood estimator $\hat{\theta}$ is calculated as

$$\hat{\text{Var}}(\hat{\theta}) = \left(\sum_{j=1}^{J} X'_{Cj}\hat{W}_{Cj}X_{Cj}n_{Cj} + \sum_{j=1}^{J} X'_{Ej}\hat{W}_{Ej}X_{Ej}n_{Ej}\right)^{-1}.$$

X_{Cj} is a vector for the control subjects in period j with $(J + 1)$ elements where entry j is equal to 1 and all other elements are equal to zero. X_{Ej} is a vector for the intervention subjects in period j with $(J + 1)$ elements where entries j and $(J + 1)$ are equal to 1 and all other elements are equal to zero. The weights $\hat{W}_{Cj}$ and $\hat{W}_{Ej}$ in the control and intervention condition are calculated as $\hat{W}_{Cj} = \hat{h}_{Cj}(1 - \hat{h}_{Cj})$ and $\hat{W}_{Ej} = \hat{h}_{Ej}(1 - \hat{h}_{Ej})$, where $\hat{h}_{Cj}$ and $\hat{h}_{Ej}$ are the hazard probabilities in the control and intervention condition in time period j.

The $\text{Var}(\hat{\beta})$, the variance of the treatment effect on the logit scale, is the element in the lower right corner of the covariance matrix $\text{Var}(\hat{\theta})$. The power level $1 - \gamma$ for the test on treatment effect with a two-sided alternative hypothesis is calculated from

$$z_{1-\gamma} \approx \frac{|\beta|}{\sqrt{\text{Var}(\hat{\beta})}} - z_{1-\alpha/2},$$

where α is the significance level and $z_{1-\gamma}$ and $z_{1-\alpha/2}$ are critical values from the standard normal distribution with probabilities $1 - \gamma$ and $1 - \alpha/2$. For a one-sided alternative hypothesis we replace $z_{1-\alpha/2}$ by $z_{1-\alpha}$.

For any design with a given total sample size, number of measurements and proportion allocation to the intervention condition the covariance matrix $\text{Var}(\hat{\theta})$ can be calculated and the $\text{Var}(\hat{\beta})$ can be used to calculate the power of the test on treatment effect. As many designs result in the same power level, the optimal design is defined as the one that is most cost-efficient. In other words, the one for which the highest power level is achieved at a certain budget, or for which the costs to achieve a desired power level are minimized.

Four cost functions are used in this paper. With the first one all subjects are followed until the end of the trial, even if they observe the event in a prior time interval. Attrition is assumed absent. The available budget is denoted C, the costs to include a subject are c_1 and the costs to take a measurement are c_2. The number of time intervals is denoted p and N is the number of subjects who enter the trial. The costs for sampling, treating, and measuring are then given by

$$C = c_1 N + c_2 N(p + 1),$$

given the assumption of equal allocation to treatment conditions. It should be noted that the number of measurements is $p + 1$ since a measurement at baseline is also taken. For all feasible number of time periods p the number of subjects N is calculated as

$$N = \frac{C}{c_1 + c_2(p + 1)}.$$

We observe that multiplying the budget C by some factor implies the sample size N for a given number of time periods p is multiplied by the same factor and the covariance matrix $\text{Var}(\hat{\theta})$ is divided by the same factor. For all designs for which the cost function holds the $\text{Var}(\hat{\beta})$ can be calculated and the one that has smallest $\text{Var}(\hat{\beta})$ is the optimal design. The efficiency of any other design, relative to the optimal one, is calculated by dividing their variances

$$\text{relative efficiency} = \frac{\text{Var}(\hat{\beta}^*)}{\text{Var}(\hat{\beta})},$$

where the $\text{Var}(\hat{\beta}^*)$ for the optimal design is always smaller than the $\text{Var}(\hat{\beta})$ for another design. Hence, the relative efficiency is always smaller than 1. As multiplying the budget C by some factor implies the $\text{Var}(\hat{\beta}^*)$ and $\text{Var}(\hat{\beta})$ are divided by the same factor the relation between the number of time periods p and relative efficiency does not change when the budget changes.

The cost function can be modified to adjust for a termination of observation after event occurrence or attrition and for unequal allocation across treatment conditions. For these modifications we refer to our earlier work (Jóźwiak & Moerbeek, 2012, 2013c).

Original Article

Optimal Design for Functional Magnetic Resonance Imaging Experiments

Methodology, Challenges, and Future Perspectives

Bärbel Maus[1,2] and Gerard J. P. van Breukelen[3]

[1]Systems and Modeling Unit, Montefiore Institute, University of Liège, Belgium, [2]Bioinformatics and Modeling, GIGA-R, University of Liège, Belgium, [3]Department of Methodology and Statistics, Faculty of Psychology and Neuroscience, Maastricht University, The Netherlands

Abstract. This paper provides an overview of optimal design for functional magnetic resonance imaging (fMRI) studies. We present the main types of fMRI designs, namely blocked and event-related designs, and common objectives of fMRI experiments, for example, localization of task-related activity in the human brain. Furthermore, we present an introduction into the methodology for analysis and optimization of fMRI experiments, for instance common analysis models and applied optimality criteria. We outline some of the problems encountered when optimizing fMRI experiments, for example, the temporal autocorrelation between measurements in fMRI data. The most important results for optimization of blocked and event-related designs with regard to the different design objectives are presented and explained. Finally, we conclude with future perspectives and challenges for optimization of fMRI experiments.

Keywords: optimal design, functional magnetic resonance imaging (fMRI), event-related design, blocked design, efficiency

Since the first functional magnetic resonance imaging (fMRI) studies in the early 1990s (Ogawa, Lee, Kay, & Tank, 1990), fMRI has provided researchers with a versatile tool to study human brain function. During an fMRI experiment the investigator typically presents stimuli, for example, images on a screen or sounds, to a subject lying in the MRI scanner, and studies the activated brain regions by statistical analysis of the collected data. The activation of brain areas causes regional changes in the blood flow and in the concentration of oxygenated blood versus deoxygenated blood which can be measured by the blood-oxygen-level-dependent (BOLD) contrast (Bandettini & Ungerleider, 2001). In this way, an indirect noninvasive measurement of neural activity is obtained (Logothetis & Wandell, 2004).

The BOLD response to a stimulus of short duration is described by the hemodynamic response function (HRF) which is characterized by a delayed onset, a subsequent rise and fall (Figure 1a). The sum of two gamma distributions called the double gamma function is often used as standard HRF in fMRI data analysis. An important finding has been the linearity of the hemodynamic response (Boynton, Engel, Glover, & Heeger, 1996). Boynton et al. (1996) showed that the hemodynamic responses to rapidly presented trials of a stimulus add up to the final observed response with only subtle deviations (Figure 1b). Furthermore, the hemodynamic response to stimuli with longer durations can be modeled approximately as sum of several individual HRFs (Boynton et al., 1996). However, nonlinearity of the HRF can occur if the interval between trials is too short, for example, less than 2 s, or for longer stimulus durations.

Optimization of experiments is especially relevant for fMRI researchers due to the low signal to noise ratio and high costs of fMRI experiments (Culham, 2006; Desmond & Glover, 2002). Culham (2006) indicates that a scanner hour can cost between 200 and 1,000 USD not including additional costs for subjects, equipment, and personnel. The signal-to-noise ratio could be increased by design replication but such a replication will further increase the high experimental costs. Optimal design is helpful in overcoming this dilemma. Resources, particularly grant funds, can be very limited in the current research environment while thought experiments, for example, including simulations or mathematical derivations, about the optimal design only require personnel costs.

Optimal design for fMRI experiments studies the efficient presentation order and sequence of stimuli during the time course of the experiment, so that, for example, the hemodynamic response in activated brain regions can be clearly differentiated from measurement noise. Efficiency means here to obtain precise parameter estimates

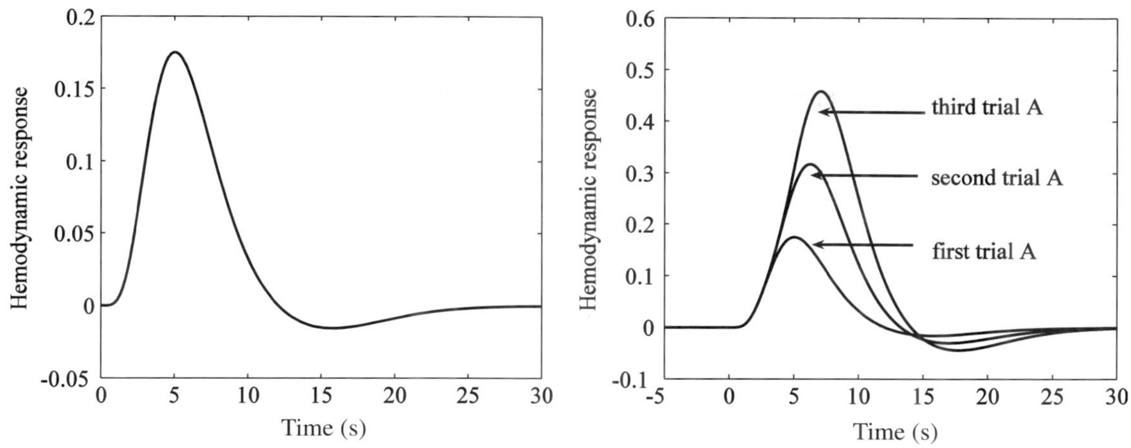

(a) HRF modelled by the standard (canonical) double gamma function

(b) Linearity of the hemodynamic response function

Figure 1. Hemodynamic response function.

with small variances or narrow confidence ellipsoid from the statistical data analysis model. In a multi-subject design, another important aspect is the choice of the number of subjects and the time length of the scanner time for each subject given a fixed total budget for the experimental costs. Increasing the number of subjects or the scan duration will always result in more information and thus more precise parameter estimation (unless there is serious bias in the additional data) but these increases will also raise the necessary study budget. It is desirable to increase efficiency within given costs or to decrease costs for a given efficiency. All these factors, presentation order, stimulus sequence, number of subjects, and experiment time length also influence the power of an experiment. At least for designs with one parameter of interest, the optimal number of subjects and scan duration for efficient parameter estimation within a given study budget will also result in a design with maximum power within this budget. In this review we focus on the optimization of the stimulus sequence presented to one subject.

While many journal papers have been written about experimental design of functional magnetic resonance imaging studies, no review paper about optimal design of fMRI experiments has been written to our knowledge for an international audience. Wüstenberg, Giesel, and Strasburger (2005) provide a general review in German about optimization of fMRI experiments but their focus is not only on experimental design. Liu (2012) focuses on the historical development and optimization of event-related designs, one major type of fMRI design. However, several book chapters have described the efficient design of fMRI experiments (Buxton, 2002; Chein & Schneider, 2003; Henson, 2007).

In this review we want to give an overview of topics and methods within optimal design research for fMRI studies. In the following section we present the most classical types of fMRI experiments and discuss how these designs compare to each other with regard to common design objectives, namely detection power, estimation efficiency, and counterbalancing. Then we describe the methodology for optimization of fMRI experiments, in particular the analysis models and optimality criteria. In the section Results for the Optimal Design of fMRI Studies we discuss known results for each type of classical fMRI experiment in more detail. In the section that follows we indicate perspectives in future research on optimization of fMRI experiments. In the final section we present a short conclusion.

Classical Design Types and Measures of Efficiency for fMRI

Blocked, Event-Related, and Mixed Designs

There are two main types of fMRI designs, blocked designs and event-related designs (see Figure 2). Furthermore in recent years, mixed designs, which are a mixture of a blocked and an event-related design have been applied (see Figure 2d). In blocked designs trials of the same stimulus type, for example, pictures of objects like houses or faces, are presented in blocks with a short interstimulus interval (ISI) between stimulus presentations. The ISI denotes the time between the end of one stimulus presentation and the beginning of the next stimulus presentation. For a design presenting pictures of houses and faces, the research question of interest can be the localization of brain areas responsible for face perception by detecting areas which are activated stronger during presentation of faces than during presentation of objects (Kanwisher, McDermott, & Chun, 1997). This corresponds to studying the contrast between the two stimulus types "faces" and "houses."

Blocks can be task blocks which consist of presentations of a certain stimulus type or simple control (reference) blocks, where the person rests or performs a passive task

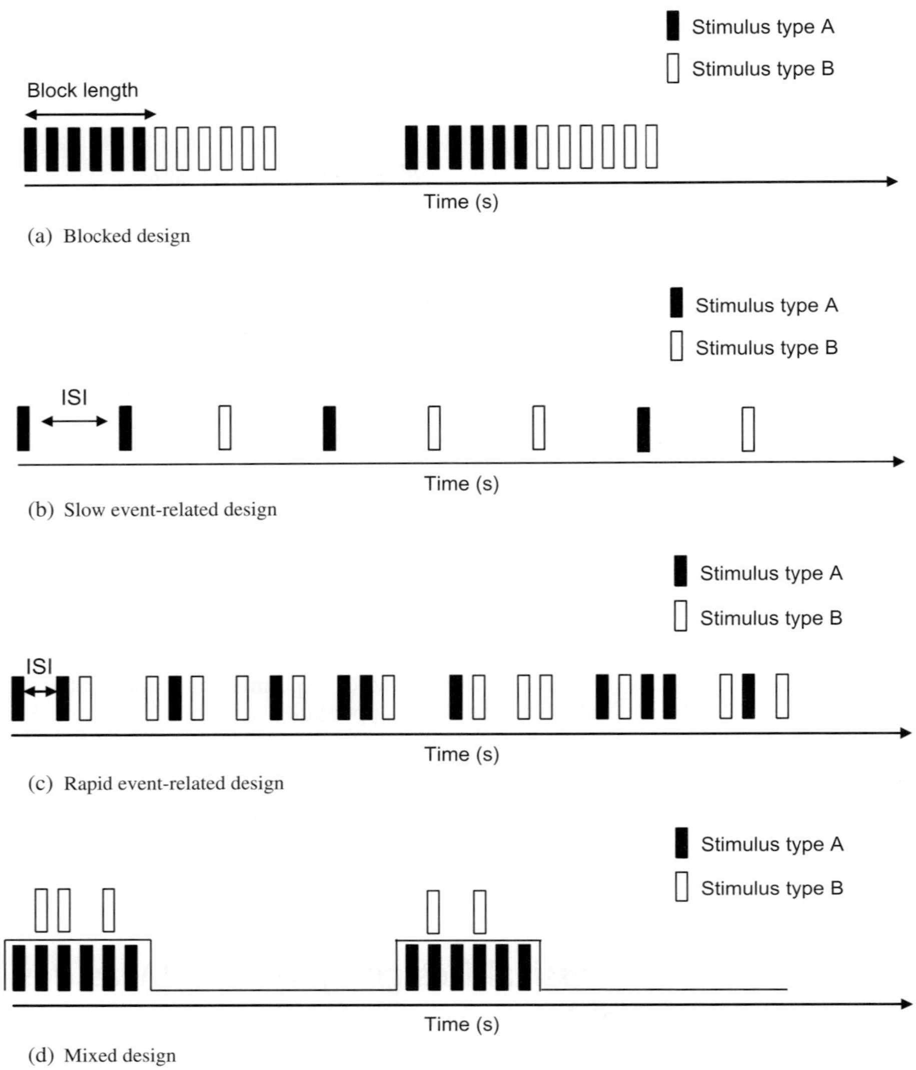

Figure 2. Main types of fMRI designs.

like watching a fixation cross. The different types of blocks are also called conditions. Two types of conditions are always needed to perform a comparison, for example, at least one task condition and one reference/control condition. The choice of an appropriate control condition is important since the comparison between the task and control condition serves to isolate the cognitive process of interest (Donaldson & Buckner, 2001). Different types of reference conditions, that is, low or high reference, are further discussed in Chein and Schneider (2003).

In the first fMRI studies, researchers used simple blocked designs with long periods of stimulus presentation as previously applied for positron emission tomography (PET) (Amaro & Barker, 2006). However, with increasing knowledge about the linearity of the HRF, event-related designs became popular. Huettel (2012) predicts that the use of blocked designs will decrease and that event-related designs will become the major type of fMRI experiments. In contrast to blocked designs, event-related designs have a random presentation order of the different stimulus types and either short randomly distributed ISIs (rapid event-related design) or a fixed long ISI (slow event-related design) between the different trials. Instead of trials one also speaks of events, hence the name event-related design. The use of randomly distributed ISIs in contrast to a fixed ISI is called jittering of ISIs.

A relatively new development are mixed designs which study sustained versus transient neural activity during task performance (Donaldson, 2004). In this design task blocks are alternated with control blocks as in a blocked design (Lazar, 2008). Additionally within task blocks trials are presented with a randomized ISI as in an event-related design (see Figure 2d). While mixed designs have been applied to study some specific cognitive processes, for example, memory, it seems that they are not yet widely used (Petersen & Dubis, 2012).

Multiple task designs, for example, cognitive conjunction designs, factorial designs with at least four conditions, and parametric designs, where the intensity of the task is modulated, have been suggested to attenuate or test the assumption of pure insertion, which postulates that one cognitive process can be added to another one without interaction between the two processes (Aguirre & D'Esposito, 1999). The "pure insertion" assumption is necessary for the approach of cognitive subtraction where the fMRI signals to two different conditions, for example, a task and control condition differing by a cognitive process of interest, are subtracted from each other to localize brain regions activated by the process of interest. Parametric, factorial, and cognitive conjunction designs can in principle be implemented as blocked or as event-related designs and are categorized by Amaro and Barker (2006) as cognitive comparison strategies. In contrast, blocked, event-related, and mixed designs are categorized by Amaro and Barker (2006) as stimulus presentation techniques. In this review we focus on optimization of stimulus presentation and present recommendations for optimal blocked and event-related designs.

Detection Power, Estimation Efficiency, and Predictability

The objectives of an fMRI experiment have been described commonly by two different measures, detection power, which is the ability to detect activated brain regions, and estimation efficiency, which is the ability to estimate the shape of the HRF. It has been shown repeatedly by simulations, theoretical equations, and geometrical explanations, that these two objectives cannot be efficiently fulfilled by the same type of design (Birn, Cox, & Bandettini, 2002; Buxton et al., 2000; Liu, 2012; Liu & Frank, 2004). Nowadays it is known that blocked designs are optimal for detection of activation, whereas rapid event-related designs are optimal for estimation of the HRF (Friston, Zarahn, Josephs, Henson, & Dale, 1999; Hagberg, Zito, Patria, & Sanes, 2001; Liu & Frank, 2004; Liu, Frank, Wong, & Buxton, 2001). However, not every randomly generated event-related design will be optimal and researchers have therefore considered different search strategies to find efficient event-related designs (Buračas & Boynton, 2002; Dale, 1999; Hagberg et al., 2001; Wager & Nichols, 2003).

It has also been shown that a design optimal for estimation of individually evoked BOLD responses, say faces or houses, might not be the optimal design for estimation of contrasts between BOLD responses (Buxton, 2002; Friston et al., 1999), for example, the contrast between the responses for viewing houses and faces to localize the brain region responsible for face recognition. There is no single optimal fMRI design and the choice of design type and its characteristics, for example, ISI, can be limited by the research question and study subjects (Chein & Schneider, 2003). To obtain an optimal trade-off between detection power and estimation efficiency, Liu and Frank (2004) and Liu (2004) considered several types of semirandom designs, for example, mixed designs, which they define as concatenations of blocked and event-related design sequences in contrast to the mixed design described in the previous section. In the section Methodology for Optimization of fMRI Experiments we present further details on optimization of blocked and event-related designs and an explanation for the trade-off between detection power and estimation efficiency.

Besides statistical efficiency as measured by detection power and estimation efficiency, psychological aspects also play a role in choosing an optimal fMRI design. Blocked designs in contrast to event-related designs with random ISI have the disadvantage of a high predictability, which can result in habituation and anticipation of the subjects so that the effect of interest will be confounded (Liu et al., 2001). The predictability of a stimulus sequence can be measured quantitatively by a design counterbalancing criterion (Kao, Mandal, Lazar, & Stufken, 2009). A counterbalanced design means that each event type is preceded and followed equally often by all other event types and is thus highly unpredictable. Designs with a high estimation efficiency typically have a low predictability, while designs with a high detection power have a high predictability (Liu, 2004). Friston et al. (1999) suggested dynamic stochastic designs which maintain some of the detection power of blocked designs while keeping some of the low predictability of random designs.

Counterbalancing the order of blocks or stimulus types in an event-related design has further the advantage of balancing out order effects for all stimulus types. Presenting a picture of a house before the picture of a face or the other way around might have an effect on the brain activity of a subject. Especially, designs with multiple conditions require careful counterbalancing to avoid confounding due to sequential and task transition effects (Chein & Schneider, 2003). Counterbalancing of block order should be performed per run or across different runs per subject and across subjects if fixed ordering is used per run, for example, A R B R A R B R ... A R B R, where A and B denote task blocks and R denotes a rest block. A run is a continuous scanning period that commonly lasts between 5 and 15 min. To combine different design objectives, for example, counterbalancing and detection of activation, multi-objective design criteria have been applied (Kao, Mandal, Lazar, et al., 2009; Wager & Nichols, 2003).

Methodology for Optimization of fMRI Experiments

The BOLD signal is sampled spatially at around hundreds of thousands of so-called voxels in the brain and temporally over one or several scanning sessions with a fixed sampling frequency, for instance, every 2 s. A scanning session is typically split into several runs. A voxel is a 3D imaging unit, for instance of the size 3 mm × 3 mm × 3 mm. A typical fMRI experiment results in an enormous amount of data with around 100,000 voxels per subject and a series of 100 to 2,000 time points per voxel (Lindquist, 2008).

In this review we will focus on the so-called mass-univariate general linear model (GLM) approach which is frequently used for the analysis of fMRI experiments. General linear models are fitted to the measured fMRI signal at each voxel separately to estimate the effects of interest taking the multiple testing problem into account, for example, via false discovery rate (Genovese, Lazar, & Nichols, 2002) or Gaussian random fields (Worsley, Taylor, Tomaiuolo, & Lerch, 2004). An overview over the multiple other methods for analysis of fMRI data can be found in Lindquist (2008).

Models for Data Analysis

During an fMRI experiment, the fMRI signal is measured repeatedly with a fixed sampling frequency called repetition time (TR) which is typically between 1 and 3 s. The number of time points for the experiment will be denoted by T. We will differentiate between two types of models, that is, a model for estimation of the HRF and a model for detection of activation.

Model for Detection of Activation

Firstly, we will describe the model which is typically used for detection of activation. Here, a HRF with fixed parameters is commonly assumed, for instance the canonical double gamma function in Figure 1a. For a given subject, the measured fMRI signal Y_i at time point i is modeled by

$$Y_i = \alpha + \beta_1 \times Z_{i1} + \beta_2 \times Z_{i2} + \cdots + \beta_Q \times Z_{iQ} + \varepsilon_i, \quad (1)$$

where the intercept α represents the baseline signal, and $\beta_1, \ldots, \beta_Q$ represent the effects of the Q stimulus types. The stimulus effects β_j ($j = 1, \ldots, Q$) correspond to the amplitude (peak height) of the hemodynamic responses for the different stimulus types and are here the parameters of interest. The expected BOLD response to stimulus type q at time point i is given by the regressors Z_{iq} (for some examples, see Figure 3e and f), which are obtained by convolution of the stimulus sequence s_q (Figure 3a and b) with the HRF (Figure 3c and d). As HRF the canonical double gamma function h given by the following equation is often used:

$$h(t) = \frac{b_1^{a_1+1} t^{a_1} e^{-b_1 t}}{\Gamma(a_1+1)} - \frac{b_2^{a_2+1} t^{a_2} e^{-b_2 t}}{c\Gamma(a_2+1)}, \quad (2)$$

where t is time in seconds (between 0 and 32 s), the parameters $a_1 = 5$, $b_1 = 1$, $c = 6$, $a_2 = 15$, and $b_2 = 1$ are fixed, and Γ denotes the gamma function. The stimulus sequence $s_q = (s_1, \ldots, s_T)$ for a given stimulus type q is a series of 0's and 1's, where 1 indicates an event of type q, for example, presentation of a stimulus, and 0 indicates no event at time point i.

The error ε is assumed to be multivariate normally distributed with expectation 0 and covariance matrix Σ. It is known that the error term is typically temporally correlated within fMRI data (Bullmore et al., 1996; Purdon & Weisskoff, 1998; Woolrich, Ripley, Brady, & Smith, 2001). To model the error within an fMRI analysis, commonly considered error structures are a first-order autoregressive (AR1) error structure (Bullmore et al., 1996) and a first-order autoregressive error plus measurement error (AR1 + ME) structure (Purdon & Weisskoff, 1998).

A first-order autoregressive error is given by a stationary variance of the error term at each time point and a decreasing correlation between errors at different time points as the time lag between these time points increases. This means that $\mathrm{Var}(\varepsilon_j) = \sigma^2$ for all $j = 1, \ldots, T$ and $\mathrm{Corr}(\varepsilon_{j_1}, \varepsilon_{j_2}) = \rho^{|j_1-j_2|}$ ($j_1, j_2 = 1, \ldots, T$), where the autocorrelation ρ is between -1 and 1. In the AR1 + ME model, the AR1 part presents correlated noise whereas the measurement error represents white (uncorrelated) scanner noise.

Model for Estimation of the Hemodynamic Response Function

Instead of assuming a certain HRF as done in model 1, we can also estimate the HRF shape. This can provide a researcher with detailed insights about brain processes and their temporal dynamics, for example, latency of responses after stimulus presentations (Wager & Nichols, 2003). For example, it could be studied whether there is a latency difference between the hemodynamic response to faces and to houses. Furthermore, it has been shown that the HRF varies between subjects and between regions in one subject (Aguirre, Zarahn, & D'Esposito, 1998; Handwerker, Ollinger, & D'Esposito, 2004; Miezin, Maccotta, Ollinger, Petersen, & Buckner, 2000). Using a misspecified HRF might decrease the power of fMRI analyses to detect activation especially for event-related designs (Aguirre et al., 1998). One main objective of fMRI experiments is thus the estimation of the HRF.

The model used for linear estimation of the HRF is similar to Equation 1 but slightly more complex. No assumptions about the shape of the HRF are considered. For simplicity, we will present the model for only one stimulus type:

$$Y_i = \alpha + h_1 \times s_i + h_2 \times s_{i-1} + \cdots + h_k \times s_{i-k+1} + \varepsilon_i. \quad (3)$$

In this model the unknown parameters of interest $h_1, \ldots, h_k$ are the k heights of the HRF during its time course, for example the $k = 33$ values of the HRF at every second over an interval of 32 s which is the typical duration of the HRF. The regressor given by the stimulus sequence s is again a series of 0's and 1's indicating an event of the stimulus type at a time point. The effective sampling rate for the HRF depends on the relationship of the employed stimulus onset asynchronies (SOAs) and the TR. The stimulus onset asynchrony denotes the time between the onset of two successive stimuli and is thus the sum of ISI and stimulus duration. A higher effective sampling rate is obtained by

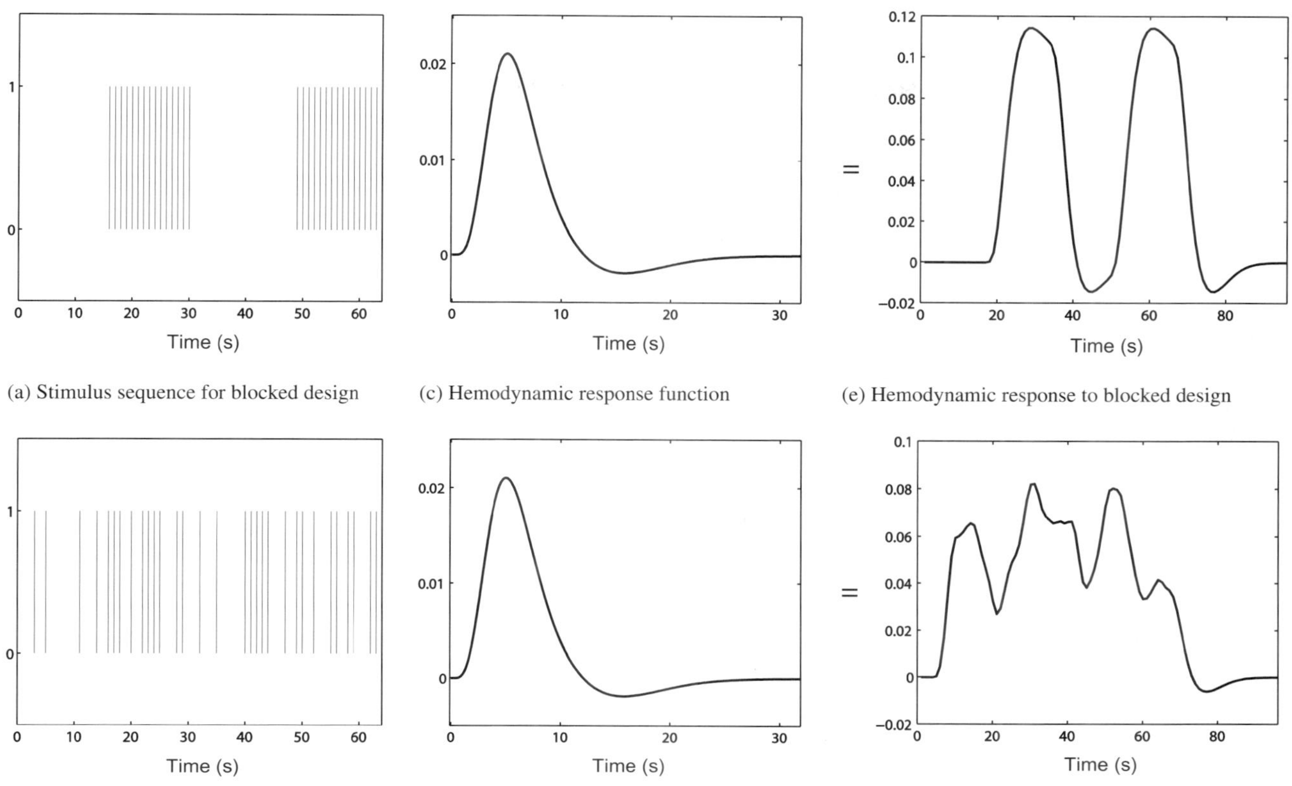

(a) Stimulus sequence for blocked design (c) Hemodynamic response function (e) Hemodynamic response to blocked design

(b) Stimulus sequence for event-related design (d) Hemodynamic response function (f) Hemodynamic response to event-related design

Figure 3. Convolution of stimulus sequences with hemodynamic response function, $\otimes$ indicates convolution.

SOAs which are not multiples of TR (Miezin et al., 2000). Please note that often ISI is used to denote SOA, and that careful reading of the literature is necessary to understand the true meaning (Garavan & Murphy, 2009).

Commonly both models in Equations 1 and 3 contain further nuisance terms besides the intercept, for example, a linear scanner drift or low frequency drifts, modeled by sine and cosine functions or lower-order Legendre polynomials. The optimization of fMRI experiments is complicated and influenced by the data's noise structure which consists of these low frequency drifts, autocorrelation, and physiological noise like heart beat and breathing. An uncorrelated error structure has often been assumed to determine analytical results for optimal fMRI experiments, but in recent years more realistic correlated structures have also been considered for optimization of fMRI studies.

Optimality Criteria

From the model equations, estimators for the parameters of interest can be derived using a generalized least squares (GLS) approach. The optimality criteria considered below focus on the covariance matrix of the GLS estimator, since this matrix reflects the accuracy of the estimator in a multidimensional space. To determine the optimal design, we will consider functions of the covariance matrix so that a one-dimensional measure of design efficiency is obtained. Different designs can then be compared to each other by their value for the optimality criterion.

In the following, we will denote the parameters of interest to be estimated with φ and their GLS estimator with $\hat{\varphi}$, for example, $\varphi = (\beta_1, \beta_2)^T$ where β_1, β_2 represent two stimulus effects as in Equation 1 for detection of activation, or $\varphi = (h_1, h_2, \ldots, h_k)^T$ where h_j represents the height of the HRF at time point j as in Equation 3 for HRF estimation, for example, $j = 1, \ldots, 17$, and the HRF is sampled over a time course from 0 to 32 s with a sampling frequency of 2 s.

Two optimality criteria have been applied for optimization of fMRI studies, the A- and the D-optimality criteria. The A-optimality criterion has been applied at first (Dale, 1999; Liu et al., 2001) and is perhaps for this reason the most popular criterion. This criterion chooses as optimal the design which minimizes the trace of the covariance matrix $\mathrm{Cov}(\hat{\varphi})$ of the parameter estimators $\hat{\varphi}$ over all possible designs. This trace is equal to the sum of the variances of the individual parameter estimators:

$$\mathrm{trace}(\mathrm{Cov}(\hat{\varphi}_1)) = \mathrm{Var}(\hat{\varphi}_1) + \cdots + \mathrm{Var}(\hat{\varphi}_p), \quad (4)$$

where p is the number of parameters of interest. The D-optimality criterion selects as optimal the design which minimizes the (pth root of the) determinant of the covariance matrix of the parameter estimators over all possible designs. For a design with two stimulus effects φ_1 and φ_2 the determinant is simply written as:

$$\det(\mathrm{Cov}(\hat{\varphi})) = \mathrm{Var}(\hat{\varphi}_1) \times \mathrm{Var}(\hat{\varphi}_2) - \mathrm{Var}(\hat{\varphi}_{12})^2, \quad (5)$$

since the off-diagonal matrix elements in $\mathrm{Cov}(\hat{\varphi})$ are equal to each other. The D-optimal design minimizes the confidence ellipsoid of the estimators, which is the multivariate analog of minimizing the width of the confidence interval for a single parameter. Commonly, nuisance effects such as linear trends and a constant baseline (intercept) are also part of the model but are not of interest. If only a subset φ_S of the estimated parameters is of interest, the A_S- or D_S-optimality criterion is applied. These criteria choose as optimal the design for which the trace or the determinant of the covariance matrix $\mathrm{Cov}(\hat{\varphi}_S)$ for the estimators of the parameter subset is minimized.

As explained before, the interest is often in estimation of contrasts between stimulus effects to detect differential activation and localize certain sub-aspects of cognitive processes, for example, by studying the contrast between "viewing faces" and "viewing houses" we can localize the brain region responsible for face recognition as distinguished from vision in general. For optimal estimation of contrasts, the C-optimality criterion can be used for which the optimal design minimizes the trace of the covariance matrix trace $(\mathrm{Cov}(C\hat{\varphi}))$ of the estimators for the stimulus effects contrasts $C\varphi$, where C is a matrix specifying the contrasts of interest (Atkinson, Donev, & Tobias, 2007). Furthermore, the generalized D-optimality criterion can be applied which aims at minimizing the determinant $\det(\mathrm{Cov}(C\hat{\varphi}))$. However, within the fMRI literature the A_S- and C-optimality are just called A-optimality criterion due to the similarity between these criteria. Likewise, the generalized D-optimality criterion and D_S-optimality criterion are just referred to as D-optimality criterion.

Local Optimality and Robust Designs

The optimal design for a given optimality criterion depends on the analysis model and on some of the unknown parameters of the model. For example, in Equation 1, the optimal design will depend on the autocorrelation parameter in an AR1 error structure which is usually unknown before conducting an fMRI experiment and might differ between individual subjects (Liu et al., 2001) and voxels (Burock & Dale, 2000). This dependence of the optimal design on given parameter values is called local optimality since a design which is optimal for certain parameter values will usually not be optimal for other values. Several methods exist to handle the problem of local optimality, for example, adaptive designs (Chow & Chang, 2008), Bayesian designs (Chaloner & Verdinelli, 1995), and maximin designs (Silvey, 1980).

For optimization of fMRI experiments, the main criterion which has been applied to date is the maximin criterion (Maus, van Breukelen, Goebel, & Berger, 2010b) to determine a design which is efficient over a range of parameter values. This criterion is based on the relative efficiency which compares one design to another design based on the applied optimality criterion. The maximin design is the design which maximizes the minimum relative efficiency when being compared to all locally optimal designs over the possible parameter space of the unknown parameters. The maximin criterion is thus a best worst-case criterion and chooses as a maximin design the design which minimizes the maximum loss of efficiency relative to the locally optimal design over a preselected region of parameter values.

In adaptive designs, results from interim analyses, which are performed while the study is running, can be used to alter the design or stop the data collection altogether. To handle local optimality, parameter estimates from an interim analysis can be used to optimize the experiment. These types of designs are becoming very popular in pharmaceutical research due to their flexibility (Chow & Chang, 2008). The Bayesian approach to handle local optimality requires the assumption of a prior distribution for the unknown parameters (Chaloner & Verdinelli, 1995; Daunizeau, Preuschoff, Friston, & Stephan, 2011). The expected efficiency of a design is then calculated by integration over this prior distribution. Examples for application of adaptive and Bayesian design optimization for fMRI experiments can be found in Xie, Clare, Gallichan, Gunn, and Jezzard (2010), Grabowski et al. (2006), and Daunizeau et al. (2011).

Multi-Objective Design Criterion

A multi-objective criterion (MO-criterion) is used to obtain an efficient design if several objectives are to be achieved by the same design, for example, high counterbalancing of trials and high detection power (Kao, Mandal, Lazar, et al., 2009; Wager & Nichols, 2003). Let F_d be the criterion measuring the detection power of a design by the A-optimality or D-optimality criterion and F_e the criterion measuring the estimation efficiency of a design by the A-optimality or D-optimality criterion. These criteria can be, for example, combined with a counterbalancing criterion F_c to avoid predictability of the design or order effects of stimuli. To formulate the multi-objective criterion, all criteria are standardized so that their values fall into the range between 0 and 1 where 0 corresponds to the least efficient possible design and 1 to the most efficient possible design. The multi-objective design criterion is then given by weighting the different design criteria relative to their importance for the researcher:

$$F = w_c \times F_c + w_d \times F_d + w_e \times F_e, \quad (6)$$

where the weights w_c, w_d, and w_e are chosen between 0 and 1 and sum up to 1, for example, $w_c = 0.5$ and $w_d = 0.5$ to combine the objectives of counterbalancing and detection power. The multi-objective optimal design is the design which maximizes the multi-objective criterion F over all possible designs.

Search Algorithm

Especially for rapid event-related designs with multiple trial types and designs combining properties of blocked and event-related designs, the space of possible designs is too huge to search it exhaustively or randomly with likely success of finding an efficient design in a reasonable time. One alternative to an exhaustive or random search is the use of one specific design which is known to be efficient, or a search within a restricted design space containing a class of efficient designs. For example, blocked designs are known to be efficient for detection power and are less variable than rapid event-related designs. However, for a complex optimization problem as given by a multi-objective criterion it can be difficult to find an optimal design via previous knowledge or a random search.

Another alternative are thus search algorithms with a more efficient strategy than an exhaustive or random search. Genetic algorithms have been proven to be a useful and flexible tool for the optimization of fMRI experiments (Kao, Mandal, Lazar, et al., 2009; Wager & Nichols, 2003). Genetic algorithms use evolutionary processes like recombination, cross-over, selection, mutation, and immigration to optimize designs over a number of generations (Holland, 1992). Wager and Nichols (2003) showed that a genetic algorithm leads to more efficient designs for detection power, estimation efficiency, and design counterbalancing than a search among randomly generated designs. Genetic algorithms are flexible, for example, with regard to the type of analysis model and design criterion, for example, detection power or estimation efficiency (Wager & Nichols, 2003). Further technical advantages of genetic algorithms for optimization of fMRI designs in comparison to other optimization techniques can be found in Wager and Nichols (2003).

Some extensions and modifications of the original genetic algorithm by Wager and Nichols (2003) have been undertaken, for example, to obtain multi-objective designs without specification of the weights for the various objectives (Kao, Mandal, & Stufken, 2012), and to account for multiple scanning sessions (Kao, Mandal, & Stufken, 2009), for nonlinear models for the HRF (Delzell et al., 2012; Maus, van Breukelen, Goebel, & Berger, 2012), and for post hoc sorting of trials depending on the subject's performance (Cordes, Herzmann, Nandy, & Curran, 2012). The trials of an experiment could, for example, be post hoc sorted depending on whether the subject correctly identified presented pictures as faces by pressing a button to indicate faces.

Results for the Optimal Design of fMRI Studies

Blocked designs and event-related designs have both their advantages and disadvantages with regard to design efficiency but also with regard to psychological aspects (e.g., attention of study participants), possible types of research questions and study populations (Aguirre & D'Esposito, 1999; Amaro & Barker, 2006; Chein & Schneider, 2003). For example, blocked designs can be used to study state-based cognitive processes such as automatic attentional processing whereas event-related designs allow for post hoc sorting of trials depending on the subjects' behavior, for example, into correct or incorrect response for further separate analysis. Here, we discuss blocked and event-related designs with regard to design efficiency as measured by optimality criteria. In the next section we focus on blocked designs whereas in the section that follows then we present event-related designs. General results for the optimal stimulus frequencies are presented in the section that follows.

Blocked Designs

In general, there are three factors which can be varied for a blocked design: block length, stimulus onset asynchrony, and block order, and we will discuss each of them below. The simplest blocked design is the so-called alternating boxcar design (Chein & Schneider, 2003) or "AB block" design with two alternating conditions A and B repeated and compared in the final analysis. To avoid confounding effects of head motion, low-frequency noise, and habituation effects (Chein & Schneider, 2003), there should always be at least two alternations (two blocks of each condition).

Optimal Block Length

For a simple alternating blocked design, where a task block alternates with a rest or control block, the optimal block length has been recommended to be between 14 and 20 s (Aguirre & D'Esposito, 1999; Chein & Schneider, 2003). This recommendation for the optimal block length can be explained by the fact that the HRF due to its shape works as a "low-pass" filter that filters out high frequency information. This low-pass filtering property of the HRF can be seen from the power spectrum of the HRF which shows higher power at lower frequencies (Aguirre & D'Esposito, 1999). As a consequence, the lower the frequency of task changes in a design, for example, the longer the block durations, the higher the detection power of this design will be (Chein & Schneider, 2003). However, the noise structure of fMRI data has to be taken into account as well for design optimization. FMRI noise is typically autocorrelated and consists of low frequency noise. To avoid a confounding effect of noise, the design frequency should thus not be within the low frequency range of the noise. This provides a trade-off between the filtering properties of the HRF favoring low frequency designs and the low frequency noise demanding high frequency designs (Aguirre & D'Esposito, 1999). Generally, block lengths should be between 10 and 60 s (Chein & Schneider, 2003). Longer block lengths should not be used due to the low frequency noise in fMRI time series and to avoid habituation and fatigue of the subjects.

For designs with two stimulus types and one possible rest condition which was also allowed to be omitted, Maus, van Breukelen, Goebel, and Berger (2010a) found slightly shorter locally optimal block lengths than commonly recommended, that is, 10 s. In particular, using the maximin approach to find designs which are robust against misspecification of the autocorrelation and variances, a maximin block length of 10 s could be recommended for the AR1 + ME model (see Table 1). Similarly for three stimulus types and one possible rest condition, Maus et al. (2010b) found maximin designs with block lengths between 10 and 14 s for detection of individual stimulus effects and contrasts between stimulus effects. These slightly shorter block lengths than normally recommended might be caused by the higher number of stimulus types than for the simple alternating AB design or by the maximin approach which aimed here at determining designs efficient over a range of possible parameter values for the autocorrelation and variance. Indeed it has been shown that the optimal block length decreases if the autocorrelation parameter increases (Maus et al., 2010b; Wager & Nichols, 2003). In the section Optimal Stimulus Frequency we will explain more about how run time should be split between the various experimental conditions.

Optimal Stimulus Onset Asynchrony

The stimulus onset asynchrony should be as short as the stimulus duration or the type of presented stimulus allow (see also Table 1). A small SOA is useful to produce a strong homogeneous signal by maximizing the number of replications within a block and to maintain the subjects' attention (Chein & Schneider, 2003). However, nonlinearities of the HRF arise for short SOAs, especially for SOAs smaller than 2 s, causing a misspecified hemodynamic response to a block of closely spaced trials. For blocked designs with equal length of blocks, nonlinearities per block will be equivalent and cause no serious bias in parameter estimation (Wager, Vazquez, Hernandez, & Noll, 2005).

Optimal Block Order

Furthermore, the block order is relevant for optimization of fMRI experiments. In the following we will denote rest blocks by R, that is, the subject rests or performs a fixation task such as viewing a fixation cross, and task blocks by A_1, A_2, …, and A_Q for the Q different stimulus/task types. If the interest is in optimizing the pairwise contrasts between several stimulus types, no rest block is needed and the optimal block order is A_1 A_2 … A_Q, which means to present one task block after the other and to repeat this order during the experiment (see also Table 1, C-optimality criterion). If however the interest is in estimation of the Q individual stimulus effects of the Q stimulus types, the statistically optimal block order is A_1 A_2 … A_Q R which means to present one task block after the other and then one rest or fixation block (see also Table 1, A_S- and D_S-optimality criterion). These results assume however that there are no

Table 1. Maximin designs for AR1 and AR1 + ME and the optimal designs for uncorrelated errors (UE; Maus et al., 2010a). Blocked designs with SOA equal to 1, 2, or 3 s, and block lengths (BL) equal to 10, 15, 20, 30, or 60 s and block orders equal to ABR, ARBR and AB were compared to each other (R = rest/fixation block, A, B = task blocks)

Optimality criteria	Error structure		
	AR1	UE	AR1 + ME
D_S	SOA = 1 s, BL = 15 s, order = ABR	SOA = 1 s, BL = 15 s, order = ABR	SOA = 1 s, BL = 10 s, order = ABR
A_S	SOA = 1 s, BL = 15 s, order = ABR	SOA = 1 s, BL = 15 s, order = ABR	SOA = 1 s, BL = 10 s, order = ABR
C	SOA = 1 s, BL = 15 s, order = AB	SOA = 1 s, BL = 20 s, order = AB	SOA = 1 s, BL = 15 s,order = AB

confounding effects due to the ordering of blocks. From a psychological viewpoint it might also be beneficial to randomize the order of the different block types. However, such a randomization has the disadvantage that the sensitivity to detect effects might be reduced since the conditions determining the contrast of most interest might be placed temporally too far away from each other to handle confounding by low frequency drift (Chein & Schneider, 2003).

Optimal Detection Power of Blocked Designs

Finally, we want to provide a simple explanation why blocked designs are optimal for detection of activation. For this purpose, one can consider again the low-pass filtering property of the HRF. Blocked designs have a consistent low frequency of task variation whereas rapid event-related designs have a wide range of design frequencies. Thus, for detection of activation where a fixed HRF is commonly assumed, blocked designs are favored in contrast to rapid event-related designs. Furthermore, in contrast to slow- or rapid event-related designs, blocked designs create a strong, homogeneous, and easy to detect signal by summation of the HRF to closely spaced trials. A more technical explanation based on a mathematical and geometrical framework can further be found in the work of Liu and colleagues (Buxton, 2002; Liu, 2012; Liu & Frank, 2004).

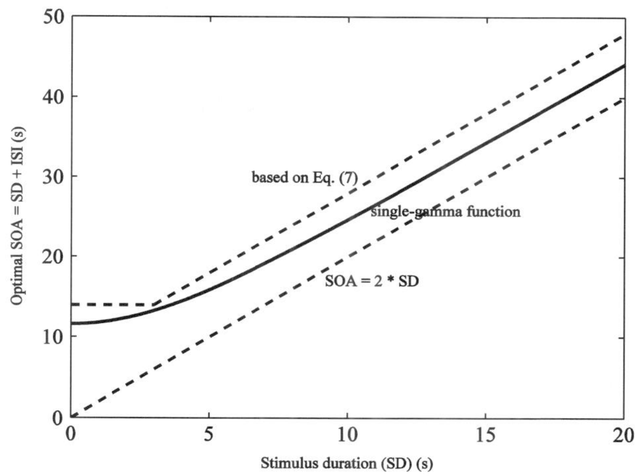

Figure 4. Optimal SOA = SD + ISI versus stimulus duration (SD) in a slow event-related design with one stimulus type. The optimal SOA is presented based on a numerical calculation assuming a single gamma distribution as HRF. For comparison the recommendation from Equation 7 and the graph for SOA = 2 × *SD* are shown (adapted from Bandettini & Cox, 2000).

Event-Related Designs

Event-related designs (or trial-based designs) are nowadays categorized into slow event-related designs and rapid event-related designs with short and randomized ISIs. Examples of a slow and rapid event-related design are given in Figure 2b and c. Slow event-related designs use typically a fixed ISI of 10 s to 16 s so that the main part of the HRF has evolved during the ISI and successive HRFs do not show substantial overlap.

Optimal Slow Event-Related Designs

The length of the optimal fixed SOA (and thus fixed ISI) in relation to a fixed stimulus duration has been a subject for optimization of slow event-related designs. Based on a theoretical calculation assuming uncorrelated errors, Bandettini and Cox (2000) present the following recommendation for the optimal SOA to detect task activation in a design with one stimulus type and stimulus duration SD (see also Figure 4):

$$\text{optimal SOA} = \begin{cases} 14 & \text{for} \quad SD < 3 \\ 14 + 2 \times (SD - 3) & \text{for} \quad SD > 3. \end{cases} \tag{7}$$

Comparison of Slow and Rapid Event-Related Designs

Birn et al. (2002) showed that estimation efficiency and detection power decrease strongly if a shorter fixed SOA than the optimal fixed SOA is used but much less if a longer fixed SOA is used (see Figure 5b and d). However, for randomized SOAs with a short mean SOA, as in rapid event-related designs, the estimation efficiency and detection power increase strongly with decreasing mean SOA at least for short stimulus durations (Birn et al., 2002; Dale, 1999; Zarahn & Friston, 2002). This is also shown in Figure 5b and d.

The comparison of event-related designs using fixed SOAs with event-related designs using short randomized SOAs has shown that the latter design has increased detection power and estimation efficiency (Birn et al., 2002) (Figure 5b and d). This higher design efficiency is caused by the fact that rapid event-related designs allow due to their randomized SOAs for faster presentation rates as well as enough variability in the hemodynamic response to detect activated regions or estimate the HRF (Miezin et al., 2000). Another disadvantage of using a fixed SOA is that the design will have a higher predictability (Friston et al., 1999). Although slow event-related designs perform thus poorly for detection power and estimation efficiency, slow event-related designs can still be useful (Chein & Schneider, 2003), for example, for estimating the HRF without the assumption of linearity for overlapping HRFs. Similarly to the study of the optimal fixed SOA in slow event-related designs, the optimal distribution of the random SOA has been studied for rapid event-related experiments.

Optimal Rapid Event-Related Designs

To generate a random distribution of the SOA (and thus fixed ISI) two parameters are often considered and varied:

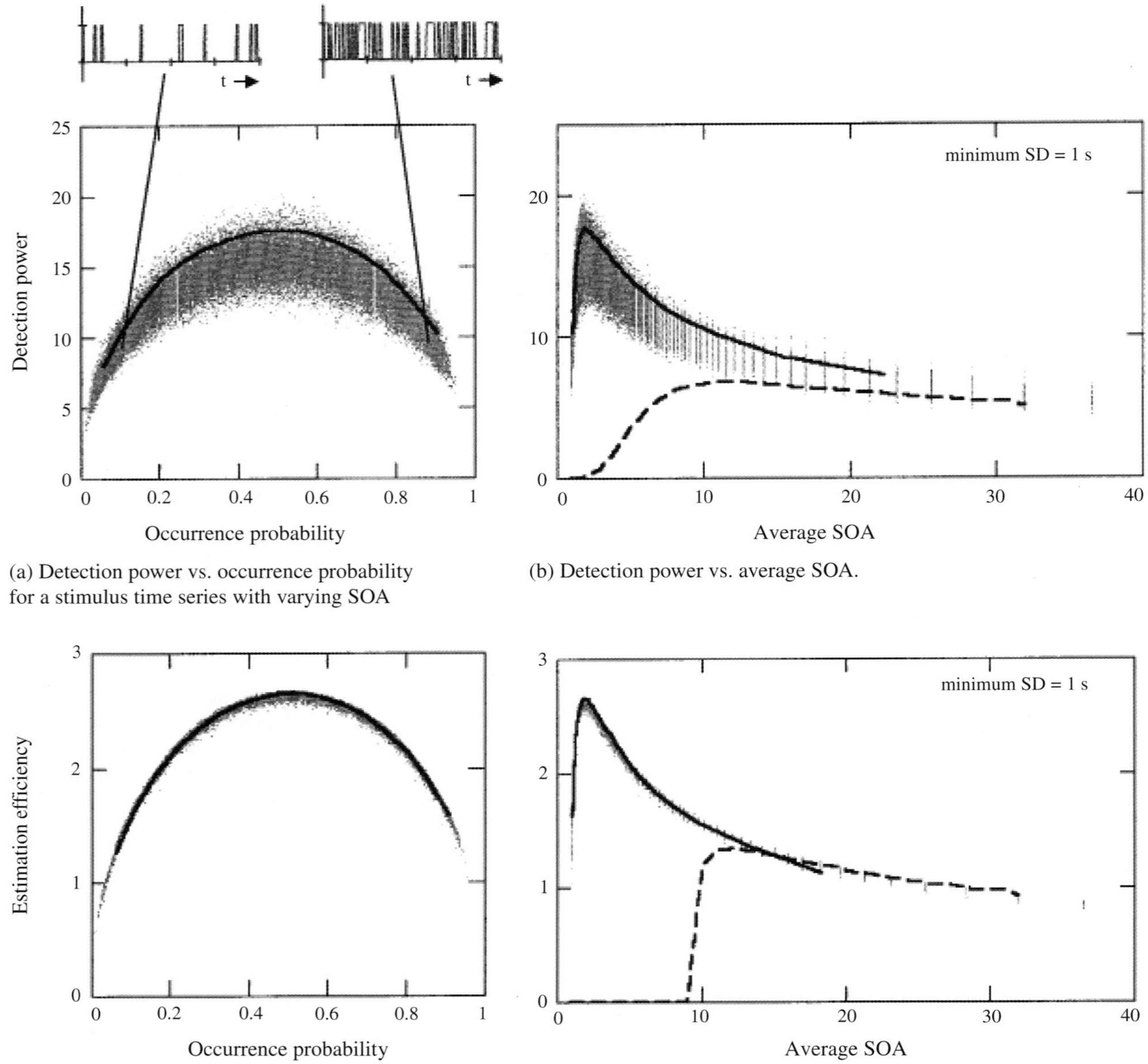

(a) Detection power vs. occurrence probability for a stimulus time series with varying SOA

(b) Detection power vs. average SOA.

(c) Estimation efficiency vs. occurrence probability for a stimulus time series with varying SOA

(d) Estimation efficiency vs. average SOA.

Figure 5. Detection power versus estimation efficiency for simulated slow and rapid event-related designs with one stimulus type. Each point represents the detection power or estimation efficiency for one simulated design. The black line represents the top 5% of stimulus patterns. The dashed line is the detectability for designs with fixed SOA. Minimum stimulus duration and TR are 1 s. The occurrence probability denotes the probability of a stimulus presentation at each time point in the experiment. The probability is fixed for a given design but is varied here between different designs. (adapted from Birn et al., 2002).

(1) the minimum SOA (SOA_{min}) between trials and (2) the occurrence probability of a stimulus type, which is the probability of that stimulus to be presented at the end of every time interval Δt, where Δt equals the minimum SOA (Friston et al., 1999). The resulting SOAs will be geometrically distributed with values which are multiples of the SOA_{min} and the trial order will be completely random. The nonoccurrence of a stimulus at a certain time point is called a "null event" and corresponds to the inclusion of a constant baseline signal in the model. For detection

power, Friston et al. (1999) showed that for an event-related design with one stimulus type the optimal occurrence probability is 0.5, meaning that stimulus and null event happen equally often every SOA_{min}, and so the optimal average SOA is given by $SOA_{min}/0.5 = 2 \times SOA_{min}$. For multiple stimulus types, Friston et al. (1999) differentiated between estimation of individual stimulus effects and contrasts between stimulus effects. For estimation of the contrast between two stimulus effects, the optimal occurrence probability is 0.5 and the shorter the SOA the better, assuming linearity of the HRF holds (Wager & Nichols, 2003). However, for estimation of two individual stimulus effects, the optimal occurrence probability is around 1/3, and so the optimal average SOA between trials of the same stimulus type is then approximately given by $3 \times SOA_{min}$. This result showed that the creation of a baseline period by null events is necessary to efficiently estimate individual responses. While a baseline period can also be achieved with longer minimum SOAs, Friston et al. (1999) also demonstrate that including null events is more efficient than increasing the minimum SOA.

Birn et al. (2002) used a slightly different approach and varied the minimum stimulus duration instead of minimum SOA in combination with variation of the occurrence probability. The authors demonstrated that estimation efficiency, that is, estimation of the HRF, is just like detection power maximized for event-related designs with one stimulus type when the occurrence probability is 0.5 (Figure 5c). Furthermore, their simulations indicated that generally detection power is increased by longer stimulus durations, causing thus more blocked like designs while estimation efficiency is decreased by longer stimulus durations. In the next section, the results of Friston et al. (1999) and Birn et al. (2002) about the optimal occurrence probability are generalized for multiple trial types.

In the following paragraph we refer the interested reader briefly to other relevant work about optimization of rapid event-related designs. The comparison of event-related designs with regard to their efficiency is complicated by the different possible ways and parameterizations to create a random distribution of the ISIs. Hagberg et al. (2001) consider different types of probability distributions to randomly distribute trails in an event-related design. One important result from Hagberg et al. (2001) is that designs with more trials are not necessarily more efficient than designs with fewer trials but that the type of distribution of trials, that is, the random distribution of ISIs, over the total run time per subject, can be more important. Friston et al. (1999) studied further also dynamic stochastic designs, where the occurrence probability of a stimulus type varies over time instead of being fixed. Buračas and Boynton (2002) demonstrated that m-sequences are highly efficient for HRF estimation since these pseudorandom sequences fulfill certain technical properties of an optimal design for estimation efficiency. Different trial orders, that is, random, alternating, and permuted (pseudorandom), were considered by Josephs and Henson (1999).

Optimal Stimulus Frequency

The optimal stimulus frequency indicates the number of time points that a certain stimulus should be presented during an experiment given the total number of time points. An important conclusion from Friston et al. (1999) is that different stimulus frequencies and SOAs are optimal for different effects of interest to be estimated. Liu and Frank (2004) present a formula for the optimal stimulus frequency using the A-optimality criterion under the assumption of uncorrelated errors, nonoverlapping stimulus types (only one stimulus presentation at each time point), the same number of trials per stimulus type and a linear trend as nuisance term in the GLM. The formula can be applied for detection power with regard to estimation of individual stimulus effects (i.e., amplitudes of the HRF) or contrasts thereof, and for estimation efficiency with regard to estimation of individual HRFs or contrasts between HRFs. Under the same previously mentioned assumptions, Maus et al. (2010b) derived the D-optimal stimulus frequency.

If the parameters of interest are pairwise contrasts between Q stimulus types or pairwise contrasts between Q HRFs, then the A-optimal stimulus frequency and D-optimal stimulus frequency are both equal to $1/Q$, meaning absence of null events. If interest is instead in the Q stimulus effects or Q HRFs themselves, then the A- and D-optimal stimulus frequencies will slightly differ. More specifically for the A-optimality criterion, the optimal stimulus frequency is then given by

$$p_A = \frac{1}{Q + \sqrt{Q}}. \tag{8}$$

The formula above implies that null events should be presented more often than trials of a certain stimulus type, at least if $Q > 1$. For one stimulus type ($Q = 1$), the optimal stimulus frequency is 0.5, which is in agreement with simulations performed by Friston et al. (1999) and Birn et al. (2002). According to Liu et al. (2001) this stimulus frequency maximizes the norm per regressor while minimizing the correlations between regressors. Such correlations between regressors inflate the standard errors of the affected coefficient estimates.

The D-optimal stimulus frequency for estimation of Q stimulus effects or Q HRFs is given by

$$p_D = \frac{1}{Q + 1}. \tag{9}$$

For the D-optimal stimulus frequencies, trials of a given stimulus type are thus presented equally often as null events. Simulations in Maus et al. (2010b) for three stimulus types indicate that the obtained optimal stimulus frequencies under the assumption of correlated errors are similar to the optimal theoretical stimulus frequencies in Equations 8 and 9 under the assumption of uncorrelated errors.

Future Perspectives

It may perhaps seem that much knowledge has already been obtained for the optimization of fMRI experiments and that hardly any further research on this topic is necessary. However, this is not the case as fMRI research is a fast-developing and versatile field with regard to statistical methodology and experimental design. To enhance the use of optimal design in the practical application of fMRI, it is necessary that statistical methodology is developed, extended and applied for more complex fMRI designs, for example, mixed designs (Petersen & Dubis, 2012), and for designs with limitations, for example, event-related designs with a fixed trial order. Furthermore, as the analysis methods for fMRI data evolve and new methods become more widely used, properties of optimal designs for these approaches need to be investigated and may lead to new types of optimal designs (Liu, 2012). Below we will outline three possible new directions for optimization of fMRI experiments, that is, optimization of mixed designs, of designs for connectivity studies (Daunizeau et al., 2011), and for multi-voxel pattern detection (Mur, Bandettini, & Kriegeskorte, 2009). Furthermore, we will consider the need to incorporate design constraints and uncertainty about the HRF shape into the design optimization. Finally, we will discuss multi-subject designs.

Design Optimization for Advanced Designs and Analysis Methods

The application of mixed designs enables a researcher to study sustained neural activity and transient activity separately by combination of a blocked design with an event-related design. Petersen and Dubis (2012) give two general recommendations for mixed designs, that is, to alternate between control and task blocks as in blocked designs and to jitter event-related trials according to a uniform distribution, but more research seems necessary. A particular issue worth studying for optimization of mixed designs are design characteristics which avoid bias between the parameter estimate for the regressor modeling the sustained activity and the parameter estimate for the regressor modeling the transient activity in the statistical analysis model (Petersen & Dubis, 2012).

Brain connectivity studies aim at discovering brain networks that describe how different brain regions interact depending on experimental conditions. Analysis methods for these types of studies have been an active area of research in recent years within the neuroimaging community and consist for example of structural equation modeling, dynamic causal modeling (DCM), and multivariate analysis methods like principal or independent component analysis. In a next step focus should be given to the optimal design for these studies (Daunizeau et al., 2011). Daunizeau et al. (2011) developed an optimality criterion to find designs which minimize the model selection error in effective connectivity studies using DCM, where model selection error means that among a set of candidate models for brain networks a model is chosen which is actually not the "closest" to the real unknown brain network. However, design optimization for connectivity studies might be complicated by the wide range of connectivity methods and their lack of standardization (Poldrack, 2012).

Another topic which has obtained much attention in recent years is multi-voxel pattern analysis (MVPA), which aims at decoding patterns of activation to a condition over multiple voxels instead of focusing on single voxel activity or regions of interest as in the classical GLM described in Equation 1 (Mur et al., 2009). For MVPA the data is commonly split into a training set for the classification algorithm and a test set to determine the performance of the obtained classification rule. To our knowledge, researchers have focused on comparing the performance of different classification algorithms and the influence of preprocessing steps on performance (Etzel, Valchev, & Keysers, 2011) but no study has evaluated in detail the characteristics of optimal experimental design specifically for MVPA. Mur et al. (2009) present some general recommendations for the choice between an event-related and blocked design for pattern analysis. Furthermore, they indicate that with regard to pattern analysis an event-related design can have increased sensitivity since it provides more independent data points than a blocked design. Coutanche and Thompson-Schill (2012) showed that several short runs resulted into increased classification accuracy in comparison to a few long runs, given the same total number of time points.

Design Constraints and Uncertainty About HRF

Another possible issue for future research on optimization of fMRI experiment is finding optimal designs within design constraints. Many fMRI experiments are limited in the choice of their design, for example, the order of the different trial types may be fixed (Liu, 2012). We can think, for example, of a lexical decision task, where the subject first sees the word "nurse" (priming) and then the word "doctor," after which the subject has to perform the decision task whether "doctor" is a word or a nonword and has to indicate his decision by a button press. An optimal design then needs to be searched in a design subspace which may not contain designs known to be optimal under less restrictive circumstances. Research about general guidelines for optimization under design constraints and about the effect of these constraints on design efficiency will be beneficial. Genetic algorithms are flexible in incorporating restrictions about possible designs and different types of models as explained by Wager and Nichols (2003), but can be slow in convergence to an optimal design. Using design constraints and more complex models may even further increase the run time of the algorithm. Therefore, the development of more time-efficient optimization algorithms is relevant (Liu, 2012). So far, genetic algorithms have been the main focus of algorithms for optimization for fMRI designs, but other algorithms might prove to be useful and be more time-efficient, while maintaining design efficiency (Kao & Mittelmann, 2013).

Another topic deserving further research is that of uncertainty about the HRF shape. Detection power has mainly been studied assuming a fixed shape of the HRF, that is, the standard double gamma function. It has however been shown that the HRF varies between subjects and regions (Aguirre et al., 1998; Miezin et al., 2000). Blocked designs are probably robust against slight misspecifications of the HRF although broader HRFs can lead to longer optimal block lengths (Liu et al., 2001; Maus, van Breukelen, Goebel, & Berger, 2011). It is unclear how different HRFs shapes affect the detection power and optimization of rapid event-related designs. The robustness of the design with regard to misspecification of the HRF is also relevant here.

Estimation efficiency has usually been studied without any prior assumption about the HRF shape. Liu and Frank (2004) study estimation efficiency when prior assumptions about the HRF shape are made, that is, two gamma distributions and their derivatives are used as basis functions to span a range of HRF shapes by linear combination. If such linear sets of basis functions to model the HRF are increasingly used, it will also be relevant to study the optimal designs for data analysis models assuming these HRF basis sets. Similarly, recent studies have studied optimal designs for estimation of the HRF by using a nonlinear regression model, where the parameters of the double gamma function are unknown and have to be estimated (Delzell et al., 2012; Maus et al., 2012). Such a nonlinear model is different from the linear regression models in Equations 1 and 3, where either the parameters of the double gamma function are assumed to be known or where no specific function at all is assumed for the HRF.

Group Studies

A last topic for further research that we want to mention here is that of multi-subject designs. Typically an fMRI experiment will consist of collecting imaging data from a given subject over time (subject level) and over several subjects (group level), so that generalizations from the experimental results to a larger population can be drawn. Design optimization at the subject level for an effect of interest in the group study will also improve parameter estimation at the group level since the error variance at the group level is composed as the sum of the variance of the effect estimator at the subject level and the between-subject variance (Mumford & Nichols, 2006). Relevant questions for optimization at the group level are the optimal number of subjects and the optimal scan duration with regard to power and design efficiency, taking into account the trade-off between the number of subjects and scan duration given a budget for the experimental costs.

Several papers have optimized multi-subject fMRI experiments by maximization of the statistical power for a given type I error (Desmond & Glover, 2002; Mumford & Nichols, 2008). Maus et al. (2011) considered the optimization of multi-subject designs from an optimal design perspective and accounted also for experimental costs. Thirion et al. (2007) have taken yet another approach and have focused on methodological aspects for group studies with regard to reliability measures for activation maps. Interestingly, several of these studies have indicated that higher numbers of subjects than the commonly applied range of 10–16 subjects for fMRI group studies would be necessary to achieve sufficient power and reliability (Thirion et al., 2007). An overview about optimization of multi-subject designs with regard to the number of subjects, number of event/scans, and scan duration is given by Garavan and Murphy (2009) who indicate that this topic deserves further attention, for example for across-subjects studies where healthy controls and clinical patients are compared.

Conclusions

Several fMRI researchers have studied the optimization of fMRI experiments, while this topic has been largely unnoticed by statisticians (Lindquist, 2008). Given that optimization of fMRI studies requires both advanced statistical understanding and practical knowledge about the complexity of fMRI experiments and their data analysis, cooperation in multidisciplinary teams of fMRI experts and statisticians would be most beneficial.

For all computer-based optimization methods, it is useful to provide an easy-to-use and robust computer tool. Unfortunately, many researchers do not profit from the developed tools and methodologies for optimization of fMRI experiments. This might be because tools are not freely available, easy to use, or do not incorporate certain design limitations. It is also relevant to show the advantages of using an optimized design versus common design strategies. Comparisons between optimized designs and applied designs might exemplify the need and raise the awareness. Kao, Majumdar, Mandal, and Stufken (2013) show that their maximin design for nonlinear estimation of the HRF increases the minimum efficiency by 34.4% in comparison to the mean efficiency of a rapid event-related designs parameterized as in Miezin et al. (2000).

In conclusion, fMRI researchers can benefit from past and future results in optimal design research for planning their study, especially if statisticians and fMRI experts increase their cooperation and design software becomes more available, flexible, and user-friendly. In this review we have provided an introduction to the most common types of designs and to the approaches for optimization of fMRI experiments. Furthermore, we have presented some of the challenges for optimization of fMRI experiments, for example, the trade-off between estimation efficiency and detection power, and the most important results for optimization of blocked and event-related designs.

References

Aguirre, G. K., & D'Esposito, M. (1999). Experimental design for brain fMRI. In C. T. W. Moonen & P. A. Bandettini (Eds.), *Functional MRI* (pp. 369–380). Berlin, Germany: Springer.

Aguirre, G. K., Zarahn, E., & D'Esposito, M. (1998). The variability of human, BOLD hemodynamic responses. *NeuroImage, 8*, 360–369.

Amaro, E. J., & Barker, G. J. (2006). Study design in fMRI: Basic principles. *Brain and Cognition, 60*, 220–232.

Atkinson, A. C., Donev, A. N., & Tobias, R. D. (2007). *Optimum experimental designs with SAS*. New York, NY: Oxford University Press.

Bandettini, P. A., & Cox, R. W. (2000). Event-related fMRI contrast when using constant interstimulus interval: Theory and experiment. *Magnetic Resonance in Medicine, 43*, 540–548.

Bandettini, P. A., & Ungerleider, L. G. (2001). From neuron to BOLD: New connections. *Nature Neuroscience, 4*, 864–866.

Birn, R. M., Cox, R. W., & Bandettini, P. A. (2002). Detection versus estimation in event-related fMRI: Choosing the optimal stimulus timing. *NeuroImage, 15*, 252–264.

Boynton, G. M., Engel, S. A., Glover, G. H., & Heeger, D. J. (1996). Linear systems analysis of functional magnetic resonance imaging in human V1. *The Journal of Neuroscience, 16*, 4207–4221.

Bullmore, E., Brammer, M., Williams, S. C., Rabe-Hesketh, S., Janot, N., David, A., ... Sham, P. (1996). Statistical methods of estimation and inference for functional MR image analysis. *Magnetic Resonance in Medicine, 35*, 261–277.

Buračas, G. T., & Boynton, G. M. (2002). Efficient design of event-related fMRI experiments using m-sequences. *NeuroImage, 16*, 801–813.

Burock, M. A., & Dale, A. M. (2000). Estimation and detection of event-related fMRI signals with temporally correlated noise: A statistically efficient and unbiased approach. *Human Brain Mapping, 11*, 249–260.

Buxton, R. B. (2002). Efficient design of BOLD experiments. In *An introduction to functional magnetic resonance imaging*. Cambridge, UK: Cambridge University Press.

Buxton, R. B., Liu, T. T., Martinez, A., Frank, L. R., Luh, W.-M., & Wong, E. C. (2000). Sorting out event-related paradigms in fMRI: The distinction between detecting an activation and estimating the hemodynamic response. *NeuroImage, 11*, S457.

Chaloner, K., & Verdinelli, I. (1995). Bayesian experimental design: A review. *Statistical Science, 10*, 273–304.

Chein, J. M., & Schneider, W. (2003). Designing effective fMRI experiments. In J. Grafman & I. H. Robertson (Eds.), *Handbook of Neuropsychology* (2nd ed., Vol. 9, pp. 299–325). Amsterdam: Elsevier.

Chow, S. C., & Chang, M. (2008). Adaptive design methods in clinical trials - a review. *Orphanet Journal of Rare Diseases, 3*, 11.

Cordes, D., Herzmann, G., Nandy, R., & Curran, T. (2012). Optimization of contrast detection power with probabilistic behavioral information. *NeuroImage, 60*, 1788–1799.

Coutanche, M. N., & Thompson-Schill, S. L. (2012). The advantage of brief fMRI acquisition runs for multi-voxel pattern detection across runs. *NeuroImage, 61*, 1113–1119.

Culham, J. C. (2006). Functional neuroimaging: Experimental design and analysis. In R. Cabeza & A. Kingstone (Eds.), *Handbook of functional neuroimaging of cognition* (2nd ed., pp. 53–82). Cambridge, MA: MIT Press.

Dale, A. M. (1999). Optimal experimental design for event-related fMRI. *Human Brain Mapping, 8*, 109–114.

Daunizeau, J., Preuschoff, K., Friston, K. J., & Stephan, K. (2011). Optimizing experimental design for comparing models of brain function. *PLoS Computational Biology, 7*, e1002280.

Delzell, D. A., Gunst, R. F., Schucany, W. R., Carmack, P. S., Lin, Q., Spence, J. S., & Haley, R. W. (2012). Key properties of D-optimal designs for event-related functional MRI experiments with application to nonlinear models. *Statistics in Medicine, 31*, 3907–3920.

Desmond, J. E., & Glover, G. H. (2002). Estimating sample size in functional MRI (fMRI) neuroimaging studies: Statistical power analysis. *Journal of Neuroscience Methods, 118*, 115–128.

Donaldson, D. I. (2004). Parsing brain activity with fMRI and mixed designs: What kind of a state is neuroimaging in? *Trends in Neurosciences, 27*, 442–444. doi: 10.1016/j.tins.2004.06.001

Donaldson, D. I., & Buckner, R. L. (2001). Effective paradigm design. In P. Jezzard, P. M. Matthews, & S. M. Smith (Eds.), *Functional MRI: An introduction to methods* (pp. 177–196). Oxford, UK: Oxford University Press.

Etzel, J. A., Valchev, N., & Keysers, C. (2011). The impact of certain methodological choices on multivariate analysis of fMRI data with support vector machines. *NeuroImage, 54*, 1159–1167.

Friston, K. J., Zarahn, E., Josephs, O., Henson, R. N. A., & Dale, A. M. (1999). Stochastic designs in event-related fMRI. *NeuroImage, 10*, 607–619.

Garavan, H., & Murphy, K. (2009). Experimental design. In M. Filippi (Ed.), *fMRI techniques and protocols* (pp. 133–149). New York, NY: Humana Press.

Genovese, C. R., Lazar, N. A., & Nichols, T. (2002). Thresholding of statistical maps in functional neuroimaging using the false discovery rate. *NeuroImage, 15*, 870–878.

Grabowski, T. J., Bauer, M. D., Foreman, D., Mehta, S., Eaton, B. L., Graves, W. W., ... Bolinger, L. (2006). Adaptive pacing of visual stimulation for fMRI studies involving overt speech. *NeuroImage, 29*, 1023–1030.

Hagberg, G. E., Zito, G., Patria, F., & Sanes, J. N. (2001). Improved detection of event-related functional MRI signals using probability functions. *NeuroImage, 14*, 1193–1205.

Handwerker, D. A., Ollinger, J. M., & D'Esposito, M. (2004). Variation of BOLD hemodynamic responses across subjects and brain regions and their effects on statistical analyses. *NeuroImage, 21*, 1639–1651.

Henson, R. (2007). Efficient experimental design for fMRI. In K. J. Friston, J. T. Ashburner, S. J. Kiebel, T. E. Nichols, & W. D. Penny (Eds.), *Statistical parametric mapping: The analysis of functional brain images* (pp. 193–210). London, UK: Academic Press.

Holland, J. H. (1992). Genetic algorithms. *Scientific American, 267*, 66–72.

Huettel, S. A. (2012). Event-related fMRI in cognition. *NeuroImage, 62*, 1152–1156.

Josephs, O., & Henson, R. N. A. (1999). Event-related functional magnetic resonance imaging: Modelling, inference and optimization. *Philosophical Transactions of the Royal Society B, 354*, 1215–1228.

Kanwisher, N., McDermott, J., & Chun, M. M. (1997). The fusiform face area: A module in human extrastriate cortex specialized for face perception. *The Journal of Neuroscience, 17*, 4302–4311.

Kao, M.-H., Majumdar, D., Mandal, A., & Stufken, J. (2013). Maximin and maximin-efficient event-related fMRI designs under a nonlinear model. *Annals of Applied Statistics*. Advance online publication. Retrieved from http://imstat.org/aoas/next_issue.html

Kao, M.-H., Mandal, A., Lazar, N., & Stufken, J. (2009). Multi-objective optimal experimental designs for event-related fMRI studies. *NeuroImage, 44*, 849–856.

Kao, M.-H., Mandal, A., & Stufken, J. (2009). Efficient designs for event-related functional magnetic resonance imaging with multiple scanning sessions. *Communications in Statistics - Theory and Methods, 38*, 3170–3182.

Kao, M.-H., Mandal, A., & Stufken, J. (2012). Constrained multi-objective designs for functional magnetic resonance imaging via a non-dominated sorting genetic algorithm. *Journal of the Royal Statistical Society Series C: Applied Statistics, 61*, 515–534.

Kao, M.-H., & Mittelmann, H. D. (2013). A fast algorithm for constructing efficient event-related fMRI designs. *Journal of Statistical Computation and Simulation*. Advance online publication. doi: 10.1080/00949655.2013.804524

Lazar, N. A. (2008). *The statistical analysis of functional MRI data*. New York, NY: Springer.

Lindquist, M. A. (2008). The statistical analysis of fMRI data. *Statistical Science, 23*, 439–464.

Liu, T. T. (2004). Efficiency, power, and entropy in event-related fMRI with multiple trial types, Part II: Design of experiments. *NeuroImage, 21*, 401–413.

Liu, T. T. (2012). The development of event-related fMRI designs. *NeuroImage, 62*, 1157–1162.

Liu, T. T., & Frank, L. R. (2004). Efficiency, power, and entropy in event-related fMRI with multiple trial types, Part I: Theory. *NeuroImage, 21*, 387–400.

Liu, T. T., Frank, L. R., Wong, E. C., & Buxton, R. B. (2001). Detection power, estimation efficiency, and predictability in event-related fMRI. *NeuroImage, 13*, 759–773.

Logothetis, N. K., & Wandell, B. A. (2004). Interpreting the BOLD signal. *Annual Review of Physiology, 66*, 735–769.

Maus, B., van Breukelen, G. J. P., Goebel, R., & Berger, M. P. F. (2010a). Optimization of blocked designs in fMRI studies. *Psychometrika, 75*, 373–390.

Maus, B., van Breukelen, G. J. P., Goebel, R., & Berger, M. P. F. (2010b). Robustness of optimal design of fMRI experiments with application of a genetic algorithm. *NeuroImage, 49*, 2433–2443.

Maus, B., van Breukelen, G. J. P., Goebel, R., & Berger, M. P. F. (2011). Optimal design of multi-subject blocked fMRI experiments. *NeuroImage, 56*, 1338–1352.

Maus, B., van Breukelen, G. J. P., Goebel, R., & Berger, M. P. F. (2012). Optimal design for nonlinear estimation of the hemodynamic response function. *Human Brain Mapping, 33*, 1253–1267.

Miezin, F. M., Maccotta, L., Ollinger, J. M., Petersen, S. E., & Buckner, R. L. (2000). Characterizing the hemodynamic response: Effects of presentation rate, sampling procedure, and the possibility of ordering brain activity based on relative timing. *NeuroImage, 11*, 735–759.

Mumford, J. A., & Nichols, T. E. (2006). Modeling and inference of multi-subject fMRI data. *IEEE Engineering in Medicine and Biology Magazine, 25*, 42–51.

Mumford, J. A., & Nichols, T. E. (2008). Power calculation for group fMRI studies accounting for arbitrary design and temporal autocorrelation. *NeuroImage, 39*, 261–268.

Mur, M., Bandettini, P. A., & Kriegeskorte, N. (2009). Revealing representational content with pattern-information fMRI–an introductory guide. *Social Cognitive & Affective Neuroscience, 4*, 101–109.

Ogawa, S., Lee, T. M., Kay, A. R., & Tank, D. W. (1990). Brain magnetic resonance imaging with contrast dependent on blood oxygenation. *Proceedings of the National Academy of Sciences of the United States of America, 87*, 9868–9872.

Petersen, S. E., & Dubis, J. W. (2012). The mixed block/event-related design. *NeuroImage, 62*, 1177–1184.

Poldrack, R. A. (2012). The future of fMRI in cognitive neuroscience. *NeuroImage, 62*, 1216–1220.

Purdon, P. L., & Weisskoff, R. M. (1998). Effect of temporal autocorrelation due to physiological noise and stimulus paradigm on voxel-level false-positive rates in fMRI. *Human Brain Mapping, 6*, 239–249.

Silvey, S. D. (1980). *Optimal design*. London, UK: Chapman and Hall.

Thirion, B., Pinel, P., Mériaux, S., Roche, A., Dehaene, S., & Poline, J.-B. (2007). Analysis of large fMRI cohort: Statistical and methodological issues for group analyses. *NeuroImage, 35*, 105–120.

Wager, T. D., & Nichols, T. E. (2003). Optimization of experimental design in fMRI: A general framework using a genetic algorithm. *NeuroImage, 18*, 293–309.

Wager, T. D., Vazquez, A., Hernandez, L., & Noll, D. C. (2005). Accounting for nonlinear BOLD effects in fMRI: Parameter estimates and a model for prediction in rapid event-related studies. *NeuroImage, 25*, 206–218.

Woolrich, M. W., Ripley, B. D., Brady, M., & Smith, S. M. (2001). Temporal autocorrelation in univariate linear modeling of FMRI data. *NeuroImage, 14*, 1370–1386.

Worsley, K. J., Taylor, J. E., Tomaiuolo, F., & Lerch, J. (2004). Unified univariate and multivariate random field theory. *NeuroImage, 23*(Suppl 1), S189–195.

Wüstenberg, T., Giesel, F. L., & Strasburger, H. (2005). Methodische Grundlagen der Optimierung funktioneller MR-Experimente [Methodological principles for optimizing functional MRI experiments]. *Radiologe, 45*, 99–112.

Xie, J., Clare, S., Gallichan, D., Gunn, R. N., & Jezzard, P. (2010). Real-time adaptive sequential design for optimal acquisition of arterial spin labeling MRI data. *Magnetic Resonance in Medicine, 64*, 203–210.

Zarahn, E., & Friston, K. J. (2002). Some limit results for efficiency in stochastic fMRI designs. *Biometrical Journal, 44*, 496–509.

Bärbel Maus

Montefiore Institute
University of Liège
Grande Traverse 10
4000 Liège
Belgium
Tel. +32 43 662774
E-mail baerbel.maus@outlook.com

Original Article

Optimal Test Assembly in Practice

The Design of the Austrian Educational Standards Assessment in Mathematics

Jörg-Tobias Kuhn[1] and Thomas Kiefer[2]

[1]Institute of Psychology, University of Münster, Germany, [2]Federal Institute for Educational Research, Innovation and Development of the Austrian School System, Salzburg, Austria

Abstract. Several techniques have been developed in recent years to generate optimal large-scale assessments (LSAs) of student achievement. These techniques often represent a blend of procedures from such diverse fields as experimental design, combinatorial optimization, particle physics, or neural networks. However, despite the theoretical advances in the field, there still exists a surprising scarcity of well-documented test designs in which all factors that have guided design decisions are explicitly and clearly communicated. This paper therefore has two goals. First, a brief summary of relevant key terms, as well as experimental designs and automated test assembly routines in LSA, is given. Second, conceptual and methodological steps in designing the assessment of the Austrian educational standards in mathematics are described in detail. The test design was generated using a two-step procedure, starting at the item block level and continuing at the item level. Initially, a partially balanced incomplete item block design was generated using simulated annealing, whereas in a second step, items were assigned to the item blocks using mixed-integer linear optimization in combination with a shadow-test approach.

Keywords: Educational assessment, large-scale assessment, optimal design, automated test assembly, simulated annealing

The main objective of most large-scale assessments (LSAs) in educational settings lies in determining what populations of students know or can do in specific content areas of interest (e.g., mathematics or reading). LSAs are either based on data from large samples (sample-based LSA), which allow generalizations to the underlying student population of interest, or on data obtained from administering an assessment to an entire student population (population-based LSA), which is often more concerned with providing information at the individual or school level. Sample-based LSAs that are mainly concerned with inferences pertaining to student populations usually cover the content area of interest very broadly, using a large number of different test items. In contrast, population-based LSAs that focus on the individual student or school level often favor a very limited number of test items (in the latter case, often a single test form is used). More well-known LSA programs include, for example, the National Assessment of Educational Progress (NAEP; U.S. Department of Education, 2003, Chapter 20) in the United States, which focuses on the assessment of mathematics and reading; the Programme for International Student Assessment (PISA; OECD, 2012), which assesses the real-life competencies in reading, mathematics, and science; or the Trends in International Mathematics and Science Study (TIMSS; Mullis, Martin, Ruddock, O'Sullivan, & Preuschoff, 2009).

This article is largely concerned with sample-based LSAs. As is the case in sample-based LSAs in educational settings, most content areas to be assessed are complex and imply a large number of relevant subdomains. Further, their mastery usually requires a range of cognitive skills or competencies. If the degree of mastery of the full content area of interest at the population level is to be assessed, then the full coverage of all relevant subdomains, skills, and competencies at the individual student level is difficult to achieve without serious problems. That is, barely any student would work through an assessment comprising several hundred items without motivation loss, test fatigue, or potentially cheating. Therefore, most sample-based LSAs provide each individual student only with a small sample of items (a "test form") that stem from a much larger pool of items. Hence, at the student level, a LSA consists of a specific combination of items representing a fraction of the content domain of interest, which is, however, adequately covered from the perspective of the total item pool. The current article focuses on aspects of optimal test design and test assembly, that is, on optimal selection and combination of items. Optimal sampling designs, although relevant in the context of LSA, will not be treated here (cf. Cochran, 1977; Mislevy, Stapleton, & Rupp, 2013).

Concerning LSA validation, broad content area coverage is an important assumption for inference in building a sound interpretive (validity) argument at the population level (Kane, 2006). Sampling both students and items has been termed multiple matrix sampling (Shoemaker, 1973; Sirotnik & Wellington, 1977). Apart from the operational and validity advantages mentioned above, multiple matrix sampling has the additional benefit of enabling a more

DOI: 10.1027/2151-2604/a000146

efficient estimation of parameter distributions in subpopulations (Lord & Novick, 1968). It should be noted, however, that the broad content coverage implied by multiple matrix sampling usually comes with relatively low individual measurement precision, in combination with a lower efficiency of item parameter estimates (Eggen & Verhelst, 2006). Multiple matrix sampling generally implies that multiple test forms are spiraled across students in a classroom in order to minimize cluster effects (Johnson, 1992).

A key question in LSA design is how to optimally select and combine items from an item pool to generate individual test forms, while simultaneously minimizing sources of error (e.g., measurement error, sampling error, equating error; Wu, 2010) and taking additional important information into account (e.g., practical design constraints like testing time, cut scores). Further, many LSAs pursue a number of separate goals (e.g., NAEP pursues such diverse goals as long-term trend, main assessment, and bridge samples, special studies, and the Trial State Assessment; Johnson, 1992), underscoring a need for a careful weighting of LSA goals that should also be reflected in the final LSA design. One important goal in LSA, for example, is trend estimation, which requires a sufficiently large number of suitable anchor items to minimize linking error (Monseur & Berezner, 2007).

Of note, in LSA, the two stages of item calibration and proficiency scaling can be distinguished. In the item calibration stage prior to the LSA, item parameters are estimated based on a sufficiently large sample, and suitable items for LSA are selected. In the next step of proficiency scaling, population parameters of interest are estimated based on a large, representative sample of students and preselected items. It is standard practice in LSA to use item response theory (IRT) for item calibration and scaling. In the following, we will refer to the process of generating different test forms and designing a LSA as test assembly (van der Linden, 1998), which has been termed booklet design elsewhere (Frey, Hartig, & Rupp, 2009; Gonzalez & Rutkowski, 2010). A test design is the result of a test assembly process. Due to the complexity of test assembly in LSAs, mathematical routines have been developed that can automatically generate test forms based on prespecified constraints, while simultaneously optimizing measurement properties or other features of interest (for an in-depth treatment, see van der Linden, 2005). This optimization process has been termed automated test assembly (ATA).

This paper has two goals. First, an overview of test assembly and ATA procedures is given, in which relevant issues of optimal design will also be discussed. Second, following the call of Frey et al. (2009), we describe in detail the test assembly process of a recent nation-wide LSA, the assessment of the Austrian educational standards in mathematics.

Test Design in LSA

Definition of Key Technical Terms

We will begin by defining some technical terms. *Items* are the tasks administered to respondents. Items are "the most basic unit of an assessment" (Gonzalez & Rutkowski, 2010, p. 129) in that each item receives a score. Items in LSAs usually have different response formats such as multiple-choice or constructed response, where the former can be scored automatically, whereas the latter usually requires trained personnel (e.g., raters for written essays, or coders for short written responses). The term *testlet* (or *item cluster*) refers to a small number of items that pertain to a common stimulus (e.g., five items that refer to the same text in a reading comprehension assessment). Items within item clusters often show high cluster-specific residual correlations that challenge the standard assumption of local independency in IRT (Wainer, Bradlow, & Wang, 2007). Items are usually drawn from an *item pool* (or *item bank*), which represents the total amount of suitable items available for use in a LSA.

An *item block* represents a number of items and/or testlets that occur in a fixed sequence. Most LSAs use item blocks for operational simplicity and to tackle carryover effects or item position effects, at least at the block level. Carryover effects imply that an item parameter of interest depends on preceding items (Frey et al., 2009). A position effect refers to the fact that the difficulty of an item depends on its position in test. For example, items tend to become more difficult if presented later in a test (Le, 2009). If uncontrolled, both carryover and item position effects threaten the IRT standard assumption of item parameter invariance, which can result in substantial calibration or scaling bias (Yen, 1980) and in biased trend estimates (Zwick, 1991).

A *test form* is the actual selection of items administered to an examinee. In LSA, test form assembly is usually based on item blocks, although item-based assembly procedures have been suggested (van der Linden, Veldkamp, & Calson, 2004). Test forms that differ in the composition or sequence of items or item blocks are considered different from each other. Most LSAs that are concerned with assessing competencies at the population level administer a number of different test forms.

Constraint is another important concept in the test assembly literature, where the term constraint implies that a property of a test satisfies a lower or upper bound (van der Linden, 2005). For example, a test developer could define that no more than 50% of items in a LSA should have a multiple choice format, or that all item difficulties (p-values) should lie between .10 and .90. In the latter case, two constraints are implied ($p \geq .10$ and $p \leq .90$). As becomes evident, constraints can be defined at different levels, that is, at the item level, testlet level, item block level, test form level, or assessment level. Further, constraints can be quantitative (e.g., number of words in a reading paragraph), categorical (e.g., an item requires either reading for gist or reading for detail as reading strategies), or logical (e.g., item A and item B cannot occur in a test form together) in nature. In contrast to a constraint, an *objective* requires some property of a test to take a minimum (or maximum) value. For example, an objective could be that the test forms used have the highest possible reliability, or that as few items as possible focus on specific assessment topics. Constraints allow a higher degree of control over the test assembly process than objectives, and should thus be preferred (van der Linden, 2005).

A key concept in test assembly is the *objective function*. An objective function is the function to be optimized in a test assembly problem. For example, in a LSA, a plausible objective function could be the IRT test information function (TIF) at one or more specific scale values of interest, for example, at the cut scores between proficiency levels of a LSA. The TIF reflects the measurement accuracy of a test. By optimizing (i.e., maximizing) the TIF at cut score scale values, classification reliability can be optimized (Ercikan & Julian, 2002). Another possible objective function pertains to parameters of subpopulation densities (e.g., latent means and variances), which are often key interest in LSA. Sometimes, multiple objective functions are of interest, and algorithms for simultaneous optimization have been developed (e.g., Veldkamp, 2002).

A final important term is *incumbent*. In iterative optimization algorithms, the incumbent refers to the current optimal solution to the design problem that is kept until a better solution is found. If no better solution can be found, the incumbent is the optimal solution.

Experimental Design in Test Assembly

In order to statistically control for unwanted sources of error (e.g., parameter bias and imprecision) in LSA, procedures from the experimental design literature like balancing, keeping constant, or blocking (e.g., Winer, Brown, & Michels, 1991) can be adapted and utilized. Because overview papers have been published concerning this issue (e.g., Frey et al., 2009; Gonzalez & Rutkowski, 2010), this paper will only briefly describe three important experimental design structures that are relevant for LSA, the balanced incomplete block design (BIB design), the partially balanced incomplete block design (pBIB design), and the generalized Youden design (e.g., Kiefer, 1975).

Prior to creating a BIB design for LSA purposes, each item to be used in a LSA is usually assigned to a fixed item block. Fixed item blocks have the advantage to statistically control for within-block carryover and item position effects, and to drastically reduce the number of possible designs in comparison to item-wise design (which may, however, be seen as a disadvantage in ATA due to imposing huge constraints, see below). The item blocks are then combined into test forms such that specific design constraints are satisfied, which aim to eliminate or minimize potential sources of error (e.g., by providing maximum efficiency of item parameter estimation and sufficient stability of test form linkage). The key constraints for a BIB design are that (a) the number of test forms in which each item block appears is fixed, (b) the number of item blocks in each test form is fixed, and (c) the frequency of each pair of item blocks across all test forms is fixed. An example BIB, consisting of seven item blocks, seven test forms, and four block positions, along with the constraint that each item block pair occurs with equal frequency (exactly twice), is shown in Table 1.

Table 1. BIB design with seven item blocks, seven test forms, and four item block positions

	Position			
Test form	1	2	3	4
1	1	4	5	6
2	2	3	4	6
3	1	2	4	7
4	3	4	5	7
5	1	2	3	5
6	2	5	6	7
7	1	3	6	7

As becomes evident, a BIB design is incomplete in the sense that not all item blocks occur in every test form. However, the BIB design in Figure 1 does not control for position effects. It is generally recommendable practice in LSA to administer every item block in all (or at least multiple) possible block positions to control both calibration and proficiency scaling for position effects. In this context, generalized Youden designs can be regarded as BIB designs that additionally control for position effects. In Table 2, a generalized Youden design based on the BIB design shown in Table 1 was constructed. As can be seen, in addition to fulfilling the BIB design constraints mentioned above, the generalized Youden design assigns each item block to each possible block position exactly once.

Generating BIB designs and generalized Youden designs is not a trivial task under regular LSA circumstances with large numbers of item blocks and test forms, because the former only exist for certain parameter combinations. A number of existing BIB and generalized Youden designs have been published (e.g., Cochran & Cox, 1957). However, in generating a LSA test design, it is also possible to elegantly integrate paradigms from experimental design into an ATA framework. The advantage of doing so is that all relevant design constraints, both structural (e.g., constraints for BIB designs mentioned above) and practical (e.g., number of easy or difficult items, number of items measuring specific reading abilities per block, optimal TIF at a prespecified cut score within item blocks), are

Table 2. Generalized Youden design with seven item blocks, seven test forms, and four item block positions

	Position			
Test form	1	2	3	4
1	6	4	5	1
2	4	3	2	6
3	1	2	4	7
4	3	5	7	4
5	5	1	3	2
6	2	7	6	5
7	7	6	1	3

made explicit and can be taken into account simultaneously. At the same time, the test design is optimized with respect to the specified objective function. However, due to specific and idiosyncratic constraints, test designs can often be only partially balanced, resulting in a pBIB design. A pBIB design is not fully balanced with respect to the parameters of interest. Instead, in a pBIB design, the BIB constraints are relaxed to provide certain bounds. For example, the frequency of each pair of item blocks across all test forms need not be the same (e.g., each pair of item blocks appears a minimum of four times and a maximum of six times in the assessment). This may raise the risk of parameter bias under some circumstances (Frey et al., 2009).

Optimal Design and ATA Methods in LSA

Optimal Design Methods in LSA

A major goal of optimal design principles (e.g., Atkinson & Donev, 1992) is to estimate statistical parameters with optimal efficiency. Optimal design methods in IRT have been outlined by Berger (1998), Buyske (2005), and Grasshoff, Holling, and Schwabe (2012). As an example, let us assume that a Rasch-scalable test should be designed that maximizes the measurement precision for an examinee with ability θ_p. The measurement precision of a test with items $i = 1, \ldots, n$ at scale value θ_p is directly proportional to its TIF, which can be defined as $I(\theta_p) = \sum_{i=1}^{n} I_i(\theta_p)$, where $I_i(\theta_p)$ represents the Fisher information function of item i at θ_p (see Lord, 1980, chap. 5). In optimal design theory, the set of all possible values (in this case, all possible item difficulties) is defined as the design space. In order to obtain an efficient solution, the design space is often discretized into well-chosen design points d, $d = 1, \ldots, D$, which in this case correspond to locations on the IRT scale of interest. Each design point is associated with a weight w_d, which can be defined in different ways (e.g., the weights can represent the proportion of items used at design point d; van der Linden, 2005). The optimal design for the test above then corresponds to the selection of items i such that $I(\theta_p) = \sum_i \sum_{d=1}^{D} w_d I_{id}(\theta_p)$ is maximized (where I_{id} represents the item information function of item i at location d). Going a step further, based on a number of prespecified scale values p, the number of items required at each of these values would be determined to optimally estimate θ_p.

However, optimal design methods have some potential drawbacks in their application to LSA. As van der Linden (2005) mentions, optimal design methods determine item parameter values that are theoretically best, whereas usually the selection of items with known parameter values is of interest. Further, optimal design methods focus on optimizing statistical properties only, whereas in LSA, often practical constraints are more important. These difficulties can be circumvented by utilizing ATA methods which are described next.

Table 3. Some typical constraints in ATA

Constraint	Mathematical formulation
Exactly 10 items for set V_1	$\Sigma_{i \in V_1} x_i = 10$
Test length 25 items	$\Sigma_i x_i = 25$
Total word count minimum 1,400	$\Sigma_i w_i x_i \geq 1{,}400$
Total word count maximum 1,500	$\Sigma_i w_i x_i \leq 1{,}500$
Items 37, 43, and 49 are mutually exclusive	$x_{37} + x_{43} + x_{49} \leq 1$
Decision variables x_i are dichotomous	$x_i \in \{0, 1\}$

ATA Frameworks in LSA

A large variety of approaches for ATA have been suggested (van der Linden, 2005). Several of these (e.g., Armstrong, Jones, & Kunce, 1998; Theunissen, 1985; van der Linden et al., 2004) include item-specific decision variables $x_i \in \{0,1\}$ which indicate whether item i is assigned to a test form ($x_i = 1$) or not ($x_i = 0$). In that way, decision variables can be used to model the objective function, and all relevant constraints can be modeled as (in)equalities in the optimization process. This approach has been termed linear programming (Nemhauser & Wolsey, 1988). Table 3 provides an overview of some typical ATA constraints and their mathematical equivalent in designing a single test form. The objective function here is to maximize the TIF at a specific value of θ_p, which would be translated to $\max y = \sum_i I_i(\theta_p) x_i$. Table 3 shows that several practical constraints can be taken into account. For example, w_i reflects the word count of item i, such that total word count in the test form lies between 1,400 and 1,500 words. Or, the amount of items from specific item sets (e.g., subdomains) can be specified. In that way, practical constraints can be actively controlled in the test form generation and production process while simultaneously optimizing statistical objectives (van der Linden & Diao, 2011).

A recommendation by van der Linden (2005) is to carefully check constraints for correctness, lack of contradictions, and feasibility. If the nature of constraints rules out all feasible solutions to an ATA problem (i.e., if no test form can be constructed that conforms to the constraints), it becomes infeasible. Under some circumstances, infeasibility can result from sequential test assembly procedures. Sequential test assembly implies that multiple test forms are constructed from an item pool by generating one full test form after another. Apart from possibly resulting in infeasibility, sequential test assembly procedures may result in assembled test forms that sequentially deteriorate in quality, because the best items are drawn from the item pool first (a problem known from the class of greedy algorithms; Nemhauser & Wolsey, 1988).

There are different solutions for solving infeasibility problems (see Timminga, 1998). One possibility is to relax constraints and perceive them as desired test properties rather than as mandatory conditions in the sense of linear programming (Swanson & Stocking, 1993). This can be achieved by introducing so-called slack variables which quantify the difference between desired and actual constraint values. For example, instead of maximizing $\sum_i I_i(\theta_p)\, x_i$, upper and lower boundaries for the target TIF at θ_p can be defined, and the weighted deviations from these boundaries can then be defined as the objective function to be minimized. In this process, specific weights can be assigned to each constraint that reflect its respective impact and importance. This procedure is known as the weighted deviations model (Swanson & Stocking, 1993).

Another solution was suggested by van der Linden (2005). The big shadow test method aims at subdividing the optimization problem into an interconnected sequence of smaller problems. In this approach, the first step in generating multiple test forms (e.g., $t = 10$ tests) consists in generating a first subset of test forms (e.g., $t = 1$ and 2) as well as a big shadow-test (e.g., representing test forms $t = 3, \ldots, 10$) that conform to specific constraints. In a next step, all items from the big shadow-test are rotated back into the item pool, and a second set of test forms (e.g., $t = 3$ and 4) as well as a big shadow test (e.g., $t = 5, \ldots, 10$) are generated. This sequence is repeated until the last set of test forms has been assembled. Additional procedures for solving infeasible problems that provide information on potential causes of infeasibility are discussed in Huitzing, Veldkamp, and Verschoor (2005).

ATA Algorithms

Several different classes of algorithms have been suggested for solving ATA optimization problems. A classical set of procedures that provide an exact optimal solution are branch-and-bound algorithms (e.g., Dakin, 1965). This class of algorithms basically operates iteratively, and starts by first computing a solution to the fully relaxed problem, which implies that all integer constraints on decision variables are removed (i.e., decision variables x_i may represent non-integer values). The value of the objective function in the fully relaxed solution represents an upper bound to the optimal solution, because the addition of a constraint (e.g., that x_i be an integer) can never improve the value of the objective function. In a next step, the integer constraint on one decision variable (e.g., x_1) is reinstated, resulting in either $x_1 = 0$ or $x_1 = 1$, implying two optimization problems. Next, the solutions to these two problems are computed. If one of the problems (e.g., for $x_1 = 1$) is infeasible, or if the value of its objective function is worse than that of the currently best feasible full integer solution (the incumbent), all solutions with a corresponding value of the decision variable can be discarded. The reason is that adding constraints in later steps neither makes infeasible problems feasible, nor can adding constraints improve the values of an objective function. If a solution obtains a better objective function value than the prior incumbent, it becomes the new incumbent. The procedure of sequentially reinstating integer values to decision variables is repeated until all solutions have been checked and are known to be infeasible, ignorable, or optimal. Branch-and-bound algorithms work very well under most typical ATA circumstances (van der Linden, 2005), and solvers using them are available as freeware (Buttrey, 2005; Diao & van der Linden, 2011).

However, in the case of an extremely large number of decision variables or highly constrained ATA problems (e.g., in constructing a BIB design), the use of branch-and-bound algorithms might be problematic (e.g., problems like memory overflow or extreme solution time demand may occur). In such cases, heuristic techniques (heuristics) are more favorable. Unlike functioning branch-and-bound algorithms, heuristics are not certain to find the optimal solution. However, if properly specified, they are very good in finding a nearly optimal solution. We next describe simulated annealing, which is part of a class of local search heuristics that have been successfully applied to ATA problems. These search heuristics are local in the sense that they imply a random walk of the optimization algorithm through the design space, and continuously compare the current solution to one or more random modifications of the current solution (i.e., solutions in the neighborhood of the current solution). The current solution is kept as the incumbent if its objective function is more favorable than that of all neighborhood solutions.

In the first step of simulated annealing (Aarts, Korst, & van Laarhoven, 1997; Veldkamp, 1999), which is based on the Metropolis algorithm (Metropolis, Rosenbluth, Rosenbluth, Teller, & Teller, 1953), an initial solution to the design problem is generated as incumbent, and the value of its objective function is calculated. In the next step, a new solution is generated by randomly changing (perturbing) the initial design. If the objective function value of the new solution is better than that of the incumbent, it becomes the new incumbent. However, in order to avoid local optimal solutions, simulated annealing also includes a (small) probability that the worse solution is kept as the incumbent. More formally, in a minimization problem, the decision rule for keeping the solution at iteration l as incumbent, compared to the incumbent at iteration $l-1$, is given by

$$\Pr(\text{acceptance}|t_k) = \begin{cases} 1 & \text{if } \mathrm{o}(l) < \mathrm{o}(l-1), \\ \exp\left[-\dfrac{\mathrm{o}(l) - \mathrm{o}(l-1)}{t_k}\right] & \text{if } \mathrm{o}(l) \geq \mathrm{o}(l-1) \end{cases}$$

where $\mathrm{o}(l)$ and $\mathrm{o}(l-1)$ are the values of the objective functions of solutions at iteration l and $l-1$, respectively. The parameter t_k, which controls the probability of accepting worse solutions than the current incumbent as new incumbent, is set to an acceptably low initial value. After the algorithm has run through a predefined maximum of L iterations, the value of t_k is slightly decreased, and l is restored to 1. That is, after L iterations, t_k is changed to $t_k = ct_{k-1}$ with $0 < c < 1$ (i.e., for each value of t_k, L iterations of the algorithm are run). Hence, the probability of accepting a worse solution as new incumbent is decreasing with each iteration of t_k. The entire process is terminated if t_k becomes smaller than an initially-defined lower bound $t_{\min}$. Simulated annealing has been used successfully in several ATA problems (e.g., Finkelman, Kim, & Roussos, 2009; Sun, Chen, Tsai, & Cheng, 2008; van der Linden et al., 2004; Verschoor, 2007).

ATA Framework in the Educational Standards Assessment in Mathematics in Austria

We now describe the ATA methods that were implemented in the Austrian nation-wide educational standards assessment in mathematics in 2012. In the remainder, we refer to this assessment as the "standards assessment."

Mathematical competency in the standards assessment was decomposed into four different content domains (e.g., numbers and measures) and four different operational domains (e.g., calculus and operations; cf. BIFIE, 2010a, 2010b). In addition, tasks could be allocated to three different complexity levels (e.g., application of basic knowledge and skills). Thus, the full competency construct in the standards assessment consisted of a three-dimensional faceted grid of 48 subdomains (or "nodes") to be assessed, where each item belonged to one content domain, one operational domain, and one complexity level, respectively. This high degree of granularity requires the item pool to contain several hundred items to validly measure the construct. However, each student is only provided with a single test form consisting only of a small subset of items.

The standards assessment pursues several different goals. A first goal is to assess mathematics competency development at the individual level, class level, and school level, respectively, which is achieved by means of a population-based LSA (based on 79,678 Austrian students in the 8th grade). That part of the assessment uses only a limited number of test items. Being a subpart of the population-based LSA, a sample-based LSA intends to estimate student achievement using a large sample of items and multiple matrix sampling. Finally, a long-term trend in competency development is modeled using data from a (sample-based) baseline study in 2009. The baseline study was the first LSA of mathematics in Austria, measuring mathematics competency as described above.

Further specific goals and conditions for the test design had to be considered, some of which were described above in more general terms. These were (1) the size of the available item pool, (2) item security, (3) longitudinal linking, (4) the competency construct, (5) testing time, and (6) potential item position and carryover effects, respectively.

Item Pool

The available item pool consisted of 270 items, of which only a few had been administered in the baseline study in 2009 (see below). The larger part of the items has been developed recently, and they were extensively piloted and calibrated in 2011 in a representative sample of students. Thus, it could be assumed that the quality of the items as well as the estimated item parameters were well known and stable. Items were also screened for differential item functioning (DIF).

Item Security

The development of high-quality items that validly represent the competency construct is costly and time consuming. In addition, the repeated administration of anchor items in the long-term trend study is a necessity rather than a desideratum. Despite administrative restraints for the handling of the test material (for test administration as well as test takers) in the population-based LSA, it was (and generally is) impossible to avoid that the content of some items becomes generally known. Therefore, those items cannot be administered further. To reduce the loss of high-quality items, as discussed above, only a smaller subset of items was administered to the students of the population-based LSA. However, that item subset was aimed to be representative of the total item pool. In contrast, the students from the sample-based LSA were provided with test forms that were assembled from the full item pool. This assessment was administered with even stronger restraints (e.g., external test administrators) than the population-based LSA, with the aim to raise test security. In addition, the administered item set was sought to be large enough to provide a stable and unbiased estimation of mathematics competency at the population level. The test forms of both the population-based LSA and the sample-based LSA were generated using equivalent criteria. By including all test forms from the population-based LSA in the sample-based LSA, both assessments could be linked.

Longitudinal Link

To investigate long-term trend, we had to compare results obtained from the earlier baseline study in 2009 to those from the current LSA. Some aspects of test administration have been changed between 2009 and the current LSA (e.g., whereas students in the baseline study were required to fill out an answer sheet, the current standards assessment requires them to provide answers directly in the test booklet), which is a possible cause for equating errors (Wu, 2010). To achieve a stable link, a bridge study was conducted. Students that were part of the bridge study were provided with the original test forms used in the baseline study (including a condition with the original administration mode, i.e., an answer sheet). Data obtained from the bridge study were used for calibration and equating purposes.

Competency Construct

The assembly of different parallel test forms was carried out such that the relevant competency domains (i.e., content areas and operational areas) were represented to a comparable degree. Also, the overall difficulties of the test forms were required to be similar. Items with different response formats were included to minimize method effects. Two economical response formats are multiple-choice and true-false (Downing, 2006). In contrast to the sample-based LSA, the population-based LSA included fewer constructed-response items, because they require an expensive recoding process.

Testing Time

The testing time in the standards assessment was 90 min. The pilot study showed that this was sufficient to administer 48 items in each test form within this time period.

Item Position Effects and Carryover Effects

We assigned the items to 15 nearly equivalent blocks, with 12 items being assigned to each block. Each test form consisted of four blocks. Each block was designed similarly, starting with easier items in an attempt to raise test motivation of the students. To account for potential item position effects at the item block level, each block (and thereby each item) was assigned twice to each possible block position in the assessment. As a result, each of the 15 blocks occurred in eight test forms. In addition, the block design took potential carryover or item position effects into account, at least at the block level. That is, each pair of blocks was constrained to occur equally often across all test forms. We also had to consider that some items were not allowed to occur in the same test form (enemy items).

ATA Method for the Block Design

The test assembly process started with the specification of a global block design (block level design), which defined constraints concerned with requirements pertaining to item position and carryover effects. In a second step (item-level design), items were allocated to the "empty" blocks from the block level design. In this section, we define the constraints for the specific pBIB that were applied in the standards assessment. We then discuss the application of different ATA methods to generate a pBIB satisfying these constraints.

Optimization Program

As discussed above, regardless of the selected ATA method, the basic optimization problem can be initially specified as a mixed-integer linear program. The ATA methods to solve these programs differ as to how relaxations of the constraints are handled. The linear program that generated a pair-balanced pBIB of 15 blocks, each of which occurring twice in any block position in the standards assessment (consisting of 30 test forms), is shown in Table 4[1].

Table 4. Constraints for the item block design in the standards assessment

Constraint	Mathematical formulation
Objective function	$\min\Sigma_{b<b'} e_{bb'}$
Key constraints	$x_{fp.} = 1$ $x_{.pb} = 2$ $x_{f.b} \leq 1$
Positional balance	$\Sigma_{f'} \leq {}_6\, x_{f'pb} = 1, b \leq 6$ $\Sigma_{f'} \geq {}_7\, x_{f'pb} = 1, b \leq 6$ $\Sigma_{f'} \leq {}_{24}\, x_{f'pb} = 1, b \geq 10$ $\Sigma_{f'} \geq {}_{25}\, x_{f'pb} = 1, b \geq 10$ $\Sigma_{f'} \geq {}_7\, x_{f'pb} = 2, b = 7, 8, 9$
Special status	$x_{f.b} = 0, f \leq 6, b \geq 7$ $x_{f.b} = 0, f \geq 25, b \leq 6$ $x_{f.1} + x_{f.7} + x_{f.14} \leq 1$
Pair balance	$x_{f.b} + x_{f.b'} - 1 \leq y_{fbb'}, b < b'$ $x_{f.b} + x_{f.b'} - 2\, y_{fbb'} \leq 0, b < b'$ $y_{.bb'} - e_{bb'} \leq 2, b < b'$ $y_{.bb'} - e_{bb'} \geq 1, b < b'$
Decision variables	$x_{fpb} \in \{0, 1\}$ $y_{fbb'} \in \{0, 1\}$ $e_{bb'} \in R^+$

$x_{fpb} \in \{0,1\}$ represents a binary decision variable that describes whether or not block b, $b \leq 15$ is in test form $f, f \leq 30$ at position $p, p \leq 4$, with $b, f, p \in N^+$. The binary variable $y_{fbb'}$ describes whether or not block b is in the test form f together with block b', and the real-valued error variable $e_{bb'}$ represents the deviance of block pair b and b' from pair balance. The entity of the decision variables will be referred to as the vector $[x, y, e]$.

To obtain the required pair-balanced block design, the decision variables were allowed to take values such that the objective function became minimal while the constraints were satisfied. Given that the constraints were satisfied, the objective function was the sum of the deviations from pair balance. The key constraints defined that exactly one block could be in every position in each test form, that every block occurred exactly twice in every available position, and that each block could occur at most once in a test form, respectively.

We further distinguished the test forms and blocks that were administered in the population-based assessment. Without loss of generality (the items were assigned to blocks in a second step), we defined the first six test forms

[1] A dot within the indexes of a decision variable implies that we mean to sum over the omitted entry. For example, $x_{f.b}$ decides whether or not block b is in *any position* of test form f. In addition, for conveniency, the constraints are valid for the entire extent of each free index. For example, $x_{f.b} \leq 1$ is to be read as $\sum_p x_{fpb} \leq 1$, for each $f \leq 30$, $b \leq 15$.

and the first six item blocks (and only those) to be provided in the population-based assessment. With this definition, the overall constraint of positional balance could be decomposed into three groups of constraints defining positional balance for both groups of blocks in both groups of test forms.

The special status of the first six as well as the last six test forms was defined as follows. The last six test forms were designed such that they contained none of the first six blocks. These test forms thereby contain only items that were administered in the sample-based assessment, such that a higher degree of item security could be assumed. These test forms are proposed to be the link for the next cycle in the standards assessment. Despite the pair balance constraint, we defined three item blocks such that no test form contained a pair of these blocks. Thereby, we provided enough space for the assignment of enemy items.

The frequency of each pair of item blocks across all test forms is defined with the help of consistency constraints that assure that the variable $y_{fbb'}$ is 1 if and only if both b and b' are in form f, that is, if and only if both $x_{f.b} = 1$ and $x_{f.b'} = 1$ (cf. van der Linden, 2005; Verschoor, 2007). We define pair balance as the minimal deviance from at least one and at most two occurrences of each block pair across all test forms.

Application of ATA Methods

If a linear optimization problem is feasible, it can be directly solved using the mixed-integer linear optimization and the branch-and-bound method as discussed above (e.g., Dakin, 1965; van der Linden, 2005). We used the freeware solver lp_solve version 5.5 (retrieved from http://lpsolve.sourceforge.net/5.5/) in combination with the flexible R framework (R Development Core Team, 2012) as described by Diao and van der Linden (2011).

However, due to the inclusion of the pair balance variables $y_{fbb'}$, and due to the inclusion of the respective constraints, the size of the design problem virtually exploded. Note that without the constraints and variables regarding pair balance, the linear program would be approximately one fourth of its original size. Within the framework outlined above, the present optimization program could not be solved in reasonable time.

Therefore, we used the simulated annealing method as discussed for block designs in van der Linden et al. (2004) and described above. The objective function, resembling the fitness function of genetic algorithms (Verschoor, 2007), was defined as $c([x, y, e]) = \sum_{b<b'} e_{bb'}$ penalized by a summand representing the constraints. For example, the violation of the first key constraint $x_{fp.} = 1$ could turn into a penalization term $g_1([x, y, e]) = \lambda(h(x_{fp.} - 1))$, with a predefined function h (e.g., $h(u) = 0$ if the constraint is satisfied, and $h(u) = |u|$ otherwise). The coefficient λ is called the penalty multiplier (also known as Lagrange multiplier). As discussed above, with the help of λ we could (and should) weight the violation of crucial to less crucial constraints accordingly. This will be especially helpful in

Table 5. Generated item block design using simulated annealing

	Position			
Test form	1	2	3	4
1	5	1	4	3
2	6	3	5	2
3	1	2	3	4
4	2	5	6	1
5	3	4	2	6
6	4	6	1	5
7	7	13	8	3
8	6	11	9	8
9	9	12	8	1
10	4	10	9	7
11	13	14	4	6
12	15	8	1	4
13	5	15	7	9
14	3	1	11	10
15	2	9	7	11
16	10	6	12	5
17	8	2	10	7
18	11	7	5	8
19	1	9	2	13
20	12	8	14	2
21	9	3	15	14
22	14	5	13	9
23	7	4	3	12
24	8	7	6	15
25	10	14	15	13
26	12	13	10	14
27	15	11	14	12
28	13	15	12	11
29	14	12	11	10
30	11	10	13	15

the definition of the optimization problem in the next section. However, the multiplier and the function h should be chosen such that the penalty terms are strictly positive. The objective function is

$$f([x, y, e]) = -c([x, y, e]) + \Sigma g([x, y, e]).$$

By setting the parameters of the simulated annealing process to $t = 1$, $t_{min} = 0.00001$, $c = 0.9$, and $L = 150{,}000$, we chose a rather conservative setting for leading the algorithm (resulting in more than 15 million perturbations). Hence, we could be more liberal in performing the perturbation step. For example, we could select test forms sequentially. For each test form, we sampled a set of four blocks and assign them to available block positions within the test form. If the sampled test form reduced the violations of the constraints, the generated design was highly likely to become the incumbent. At the end of this process, the block design in Table 5 emerged.

Note that the first six forms were administered in the population-based assessment as well as in the sample-based assessment. Also, the first six forms (as well as the last six forms) resemble a pBIB design with six blocks, six forms, and four block positions. Further note that none of the blocks contained within the first six forms were contained within the last six forms. Therefore, these last six forms only consist of item blocks that are administered with stronger security restraints. Thus, these forms may be administered in the next cycle of the standards assessment.

ATA Method for the Item Block Assembly

Having generated an "empty" block design, we next filled the blocks with items. We did not need to explicitly model requirements for balancing item position and carryover effects, because the block design had taken those into account. In this section, we define the constraints for the specific item block assignments that take into consideration the requirements discussed above. We then discuss the application of a selected ATA method to generate an item block assignment that satisfies these constraints.

Optimization Problem

The linear program used to assign items to item blocks such that the requirements from the test framework were satisfied is shown in Table 6.

For this linear program, $x_{bi} \in \{0,1\}$ was the binary decision variable deciding whether or not item i, $i \leq 270$ is in block b, $b \leq 15$, with $i, b \in N^+$. Some constraints were defined at the test form level, that is, we sum over all relevant x_{bi} where b is a block in a respective test form. If not stated otherwise, the constraints were defined for the entire extent of all free indexes. Another decision variable was the error variable e_{bp} storing the deviance from the desired cumulated test information value of block b at competency node p. The entity of the decision variables is referred to as $[x, e]$.

Table 6. Constraints for assigning items to blocks using linear programing

Constraint	Mathematical formulation
Objective function: cumulated deviance from target TIF	$\min\Sigma_{b,p}\, e_{bp}$ $\Sigma_i\, (I_i(\theta_p)x_{bi}) + e_{bp} \geq t_p$
Key constraints	$x_{b\cdot} = 12$ $\Sigma_{b\in \text{Form}\, f}\, x_{bi} \leq 1$
Enemy items	$\Sigma_{b\in \text{Form}\, f}\, \Sigma_{i\in Venemy}\, x_{bi} \leq 1$
Construct domains	$\Sigma_{i\in Vcontent}\, x_{bi} \geq 3$ $\Sigma_{i\in Voperational}\, x_{bi} \geq 3$ $\Sigma_{i\in Vcomplexity}\, x_{bi} \geq 4$
Constructed-response items	$\Sigma_{b\in \text{Form}\, f}\, \Sigma_{i\in Vcr}\, x_{bi} \leq 3, f \leq 6$ $\Sigma_{b\in \text{Form}\, f}\, \Sigma_{i\in Vcr}\, x_{bi} \leq 4, f \geq 7$
High-quality items	$\Sigma_{b\in \text{Form}\, f}\, \Sigma_{i\notin \text{Vquality}}\, x_{bi} = 0, f \leq 6$ $\Sigma_{b\in \text{Form}\, f}\, \Sigma_{i\notin \text{Vquality}}\, x_{bi} = 0, f \geq 25$
Decision variables	$x_{bi} \in \{0, 1\}$ $e_{bp} \in R^+$

The optimization program aimed at implementing all relevant constraints in the test design of the standards assessment, as discussed above. Thus, there were some key constraints that had to be satisfied for consistency reasons. Each block should consist of 12 items, whereas no item was allowed to be in multiple blocks of any test form. Also, at most one of a group of enemy items was allowed to occur in any of the test forms. To balance construct representation within each test form and each block, constraints regarding construct domains had to be satisfied. That is, we wanted at least three items of each content area and of each operational area, respectively, to occur in block b. In addition, we wanted at least four items of each complexity level to be present in block b. Note that there are four content areas; with three items each in a block, the 12 items in this block are mixed; thus, the inequalities could be strengthened to an equality restriction. The costly recoding process was accounted for via the constraints regarding the number of constructed-response items. In each of the population-based LSA test forms one to six, a constraint of at most three constructed-response items was defined. For the sample-based LSA forms, however, we allowed a third of the items to have this response format.

We also specified constraints regarding items of highest quality, which we wanted to be contained only in the test forms of the population-based LSA and the potential linking forms. We specified several scale values such that the TIF for each of these values exceeded a lower bound in all blocks. The scale values were based on results from a standard setting (Freunberger & Yanagida, 2012) to enable the most reliable assignment of students to one of four competency levels.

Application of ATA Methods

If a linear optimization problem is feasible, it can be directly solved using the mixed-integer linear optimization as discussed above. However, the time bounds described for the block design also apply to the item-level design. To obtain an efficient solution to the program, we first relaxed some constraints using slack variables. Before, we needed to identify structurally critical constraints, such as the key constraints and the constraints regarding enemy and high-quality items. These constraints had to be satisfied under all circumstances. Some of the other constraints, however, might allow violations to some extent. Among these constraints were those regarding the homogeneity of the construct domain within each block. The introduction of slack variables $\gamma_{bc} \in R^+$ for a constraint group c resulted in an additional penalty term in the objective function. For example, relaxing the fourth group of constraints $\Sigma_{i\,\in\,\text{Vcontent}}\, x_{bi} \geq 3$ defined the replacing constraining group $\Sigma_{i\,\in\,\text{Vcontent}}\, (x_{bi}) + \gamma_{b4} \geq 3$. The slack variable added a penalty term to the objective function $\min\Sigma_{b,p}\, e_{bp} + \lambda_4\Sigma_b\, \gamma_{b4}$. The penalty multiplier could be used to rank the constraints by relevance and put weight on the violations of the constraints accordingly.

Secondly, we used a version of the big shadow-test method (van der Linden, 2005) in order to counterbalance the greedy character of sequential assembly procedures (see also Ariel, Veldkamp, & van der Linden, 2004, for a related approach). We assigned items to blocks sequentially, starting with the first block. Simultaneously, using an analogous set of constraints, we assigned items to "shadow blocks" yet unfilled to prevent bankruptcy of the item pool. We also ranked the blocks based on their importance with respect to the test framework, that is, we started with the first six blocks and the last six blocks, respectively. Using this method, we succeeded in decomposing a relatively large optimization program, and an adequate solution could be obtained in reasonable time. All constraints mentioned in Table 6 were met in the final design.

Concluding Remarks

In LSA, test design is a key factor in psychometric procedures like parameter estimation, scaling trend estimation, and equating, and consequently also in the interpretation and use of test scores and results. In light of this and the cost incurred by conducting a LSA, it is of utmost importance to create the most optimal test design that can be obtained. Recent years have seen a substantial growth of scientific literature pertaining to subjects like optimal test assembly and ATA, and a more widespread use of these algorithms in testing programs can be expected.

We want to conclude the paper by highlighting three important points. First, whereas we have mainly focused on ATA procedures given an existing item pool, ATA methods can already be used in prior stages, for example, in the stage of item writing (Ariel, van der Linden, & Veldkamp, 2006). This has the advantage that a balanced item pool can be obtained even before ATA algorithms for LSA are implemented, thus enlarging the design space and, hence, the feasibility of an optimal solution.

Second, the ATA procedure described in this paper represents a two-stage approach, where an optimal block design is generated first, and optimal item-to-block-allocation is conducted as a second step. Advantages like block-wise control of carryover or item position effects notwithstanding, this procedure implies severe constraints on the design space. Using ATA procedures, it is easily possible to generate test forms that conform to BIB design constraints by directly assigning single items to test forms, which may potentially boost the objective function of interest. However, this procedure allows less strict control over carryover or item position effects. In the end, as mentioned by van der Linden et al. (2004), it all boils down to how the relationship between these two factors (carryover or item position effects vs. likely higher objective function) and their relative impact is assessed.

Third, item parameters were treated as fixed in the present study, without taking uncertainty in parameter estimation into account. However, uncertainty in item parameter estimates, especially in the discrimination parameter, can result in substantial overestimation of the TIF in ATA-generated test forms (Veldkamp, Matteucci, & de Jong, 2013). Although in the LSA described in this paper, a substantial TIF overestimation is unlikely due to several reasons (e.g., by utilization of the Rasch model for scaling, usage of more than half of the item pool in the LSA), the consideration of parameter uncertainties in ATA is a promising venue of further research.

References

Aarts, E. H. L., Korst, J. H. M., & van Laarhoven, P. J. M. (1997). Simulated annealing. In E. Aarts & J. K. Kenstra (Eds.), *Local search in combinatorial optimization* (pp. 91–120). New York, NY: Wiley.

Ariel, A., van der Linden, W. J., & Veldkamp, B. P. (2006). A strategy for optimizing item-pool management. *Journal of Educational Measurement, 43*, 85–96.

Ariel, A., Veldkamp, B. P., & van der Linden, W. J. (2004). Constructing rotating item pools for constrained adaptive testing. *Journal of Educational Measurement, 41*, 345–359.

Armstrong, R. D., Jones, D. H., & Kunce, C. S. (1998). IRT test assembly using network-flow programming. *Applied Psychological Measurement, 22*, 237–247.

Atkinson, A. C., & Donev, A. N. (1992). *Optimum experimental designs*. Oxford, UK: Oxford University Press.

Berger, M. P. F. (1998). Optimal design of tests with dichotomous and polytomous items. *Applied Psychological Measurement, 22*, 248–258.

BIFIE. (Ed.). (2010a). *Bildungsstandards in Österreich: Überprüfung und Rückmeldung* [Educational standards in Austria: Assessment and feedback]. Salzburg, Austria: Bundesinstitut für Bildungsforschung, Innovation & Entwicklung des österreichischen Schulwesens.

BIFIE. (Ed.). (2010b). *Praxishandbuch für „Mathematik" 8. Schulstufe* [Practical handbook for mathematics in grade 8]. Graz, Austria: Leykam.

Buttrey, S. E. (2005). Calling the lp_Solve linear program software from R, S-PLUS and Excel. *Journal of Statistical Software, 14*, 1–13.

Buyske, S. (2005). Optimal design in educational testing. In M. P. F. Berger & W. K. Wong (Eds.), *Applied optimal designs* (pp. 1–19). New York, NY: Wiley.

Cochran, W. G. (1977). *Sampling techniques* (3rd ed.). New York, NY: Wiley.

Cochran, W. G., & Cox, G. M. (1957). *Experimental designs* (2nd ed.). New York, NY: Wiley.

Dakin, R. J. (1965). A tree-search algorithm for mixed integer programming problems. *The Computer Journal, 8*, 250–255.

Diao, Q., & van der Linden, W. J (2011). Automated test assembly using lp_Solve Version 5.5 in R. *Applied Psychological Measurement, 35*, 398–409.

Downing, S. M. (2006). Selected-response item formats in test development. In S. M. Downing & T. M. Haladyna (Eds.), *Handbook of test development* (pp. 287–301). Mahwah, NJ: Erlbaum.

Eggen, T. J. H. M., & Verhelst, N. D. (2006). Loss of information in estimating item parameters in incomplete designs. *Psychometrika, 71*, 303–322.

Ercikan, K., & Julian, M. (2002). Classification accuracy of assigning student performance to proficiency levels: Guidelines for assessment design. *Applied Measurement in Education, 15*, 269–294.

Finkelman, M., Kim, W., & Roussos, L. A. (2009). Automated test assembly for cognitive diagnosis. *Journal of Educational Measurement, 46*, 273–292.

Freunberger, R., & Yanagida, T. (2012). Kompetenzdiagnostik in Österreich: Der Prozess des Standard-Settings [Competency diagnostics in Austria: The standard-setting process]. *Psychologie in Österreich, 5*, 396–403.

Frey, A., Hartig, J., & Rupp, A. A. (2009). An NCME instructional module on booklet designs in large-scale assessments of student achievement: Theory and practice. *Educational Measurement: Issues and Practice, 28*, 39–53.

Gonzalez, E., & Rutkowski, L. (2010). Principles of multiple matrix booklet designs and parameter recovery in large-scale assessments. In M. von Davier & D. Hastedt (Eds.), *IERI monograph series: Issues and methodologies in large-scale assessments* (Vol. 3, pp. 125–156). Hamburg, Germany: IEA-ETS Research Institute.

Grasshoff, U., Holling, H., & Schwabe, R. (2012). Optimal designs for the Rasch model. *Psychometrika, 77*, 710–723.

Huitzing, H. A., Veldkamp, B. P., & Verschoor, A. J. (2005). Infeasibility in automated test assembly models: A comparison study of different methods. *Journal of Educational Measurement, 42*, 223–243.

Johnson, E. G. (1992). The design of the National Assessment of Educational Progress. *Journal of Educational Measurement, 29*, 95–110.

Kane, M. T. (2006). Validation. In R. L. Brennan (Ed.), *Educational measurement* (pp. 17–64). Westport, CT: Praeger.

Kiefer, J. (1975). Balanced block designs and generalized Youden designs. I. Construction (patchwork). *Annals of Statistics, 3*, 109–118.

Le, L. T. (2009). Effects of item positions on their difficulty and discrimination – A study in PISA science data across test language and countries. In A. Okada, T. Maizumi, & T. Hoshino (Eds.), *New trends in psychometrics* (pp. 217–226). Tokyo, Japan: Universal Academic.

Lord, F. M. (1980). *Applications of item response theory to practical testing problems*. Hillsdale, NJ: Erlbaum.

Lord, F. M., & Novick, M. R. (1968). *Statistical theories of mental test scores*. Reading, MA: Addison-Wesley.

Metropolis, N., Rosenbluth, A. W., Rosenbluth, M. N., Teller, A. H., & Teller, E. (1953). Equation of state calculations by fast computing machines. *The Journal of Chemical Physics, 21*, 1087–1092.

Mislevy, J. L., Stapleton, L. M., & Rupp, A. A. (2013). Complex person and item sampling: Implications for analysis and inference. In M. Simon, K. Ercikan, & M. Rousseau (Eds.), *Improving large-scale assessment in education: Theory, issues, and practice* (pp. 207–236). New York, NY: Routledge.

Monseur, C., & Berezner, A. (2007). The computation of equating errors in international surveys in education. *Journal of Applied Measurement, 8*, 323–335.

Mullis, I. V., Martin, M. O., Ruddock, G. J., O'Sullivan, C. Y., & Preuschoff, C. (2009). *TIMSS 2011 assessment frameworks*. Chestnut Hill, MA: TIMSS & PIRLS International Study Center.

Nemhauser, G. L., & Wolsey, L. A. (1988). *Integer and combinatorial optimization*. New York, NY: Wiley.

OECD. (2012). *PISA 2009 technical report*. Paris, France: OECD.

R Development Core Team. (2012). *R: A language and environment for statistical computing [Computer software manual]*. Vienna, Austria.

Shoemaker, D. M. (1973). *Principles and procedures of multiple matrix sampling*. Cambridge, MA: Ballinger.

Sirotnik, K., & Wellington, R. (1977). Incidence sampling: An integrated theory for "matrix sampling". *Journal of Educational Measurement, 14*, 343–399.

Sun, K.-T., Chen, Y.-J., Tsai, S.-Y., & Cheng, C.-F. (2008). Creating IRT-based parallel test forms using the genetic algorithm method. *Applied Measurement in Education, 21*, 141–161.

Swanson, L., & Stocking, M. L. (1993). A model and heuristic for solving very large item selection problems. *Applied Psychological Measurement, 17*, 151–166.

Theunissen, T. (1985). Binary programming and test design. *Psychometrika, 50*, 411–420.

Timminga, E. (1998). Solving infeasibility problems in computerized test assembly. *Applied Psychological Measurement, 22*, 280–291.

U.S. Department of Education. (2003). *NCES handbook of survey methods*. Washington, DC: Author.

van der Linden, W. J. (1998). Optimal assembly of psychological and educational tests. *Applied Psychological Measurement, 22*, 195–211.

van der Linden, W. J. (2005). *Linear models for optimal test design*. New York, NY: Springer.

van der Linden, W. J., & Diao, Q. (2011). Automated test-form generation. *Journal of Educational Measurement, 48*, 206–222.

van der Linden, W. J., Veldkamp, B. P., & Carlson, J. E. (2004). Optimizing balanced incomplete block designs for educational assessments. *Applied Psychological Measurement, 28*, 317–331.

Veldkamp, B. P. (1999). Multiple objective test assembly problems. *Journal of Educational Measurement, 36*, 253–266.

Veldkamp, B. P. (2002). Multidimensional constrained test assembly. *Applied Psychological Measurement, 26*, 133–146.

Veldkamp, B. P., Matteucci, M., & de Jong, M. G. (2013). Uncertainties in the item parameter estimates and robust automated test assembly. *Applied Psychological Measurement, 37*, 123–139.

Verschoor, A. J. (2007). *Genetic algorithms for automated test assembly*. Unpublished doctoral dissertation, University of Twente, The Netherlands.

Wainer, H., Bradlow, E. T., & Wang, X. (2007). *Testlet response theory and its applications*. New York, NY: Oxford University Press.

Winer, B. J., Brown, D. R., & Michels, K. M. (1991). *Statistical principles in experimental design* (3rd ed.). New York, NY: McGraw-Hill.

Wu, M. (2010). Measurement, sampling, and equating errors in large-scale assessments. *Educational Measurement: Issues and Practice, 29*, 15–27.

Yen, W. M. (1980). The extent, causes and importance of context effects on item parameters for two latent trait models. *Journal of Educational Measurement, 17*, 297–311.

Zwick, R. (1991). Effects of item order and context on estimation of NAEP reading proficiency. *Educational Measurement: Issues and Practice, 10*, 10–16.

Jörg-Tobias Kuhn

Institute of Psychology
University of Münster
Fliednerstr. 21
48149 Münster
Germany
Tel. +49 251 8334127
Fax +49 251 8339469
E-mail t.kuhn@uni-muenster.de

Call for Papers

"Destructive Leadership – Antecedents and Consequences"

A Topical Issue of the *Zeitschrift für Psychologie*

Guest Editors: Jan Schilling (University of Applied Administrative Sciences, Hannover, Germany) and Birgit Schyns (Durham University Business School, United Kingdom)

While a lot of leadership research has focused on positive behavior, more and more knowledge has become available on negative aspects of leadership. The reason for this shift is an acknowledgment that negative leadership can have serious consequences for followers (e.g., in terms of stress, well-being, and health), organizations (such as lower performance, reputation issues), and leaders themselves (e.g., for their career path).

Negative aspects of leadership have been captured under many headings, such as destructive leadership, toxic leadership, or abusive leadership. This topical issue does not narrow the focus to one specific definition; however, defining which type of leadership is the focus of any specific contribution will be crucial so that its place in the literature can be clearly determined. Therefore, we will put particular emphasis on theoretical definitions in each contribution.

The focus of this topical issue is on understanding more about the triggers of destructive leadership, both in terms of actual destructive behavior (such as the relevance of leader personality or organizational/cultural context) as well as outcomes, specifically in how far effects of destructive leadership can be prevented or what facilitates/hinders negative outcomes of destructive leadership. We are looking for original empirical articles as well as review-type articles or theoretical contributions that outline new models of thinking about destructive leadership.

How to submit: Interested authors should submit a letter of intent including: (1) a working title for the manuscript, (2) names, affiliations, and contact information for all authors, and (3) an abstract detailing the content of the proposed manuscript to either of the guest editors:

Jan Schilling (jan.schilling@nds-sti.de)
Birgit Schyns (birgit.schyns@durham.ac.uk)

There is a *two-stage submissions process*. Initially, authors are requested to submit only abstracts of their proposed papers. Authors of the selected abstracts will then be invited to submit full papers. All papers will undergo blind peer review.

Deadline for submission of abstracts is October 15, 2013.
Deadline for submission of full papers is February 15, 2014.

The journal seeks to maintain a short turnaround time, with the final version of the accepted papers being due by June 15, 2014. The topical issue will be published as issue 4 (2014).

For additional information, please contact:

Jan Schilling (jan.schilling@nds-sti.de)
Birgit Schyns (birgit.schyns@durham.ac.uk)

General Information About the Journal

The *Zeitschrift für Psychologie* was founded in 1890 and is the second oldest psychology journal in the world. One of the founding editors was Herrmann Ebbinghaus. Since 2007, it is published in English and is devoted to publishing topical issues that provide state-of-the-art reviews of current research in psychology.

For detailed author guidelines, please see the journal's website at www.hogrefe.com/journals/zfp/

DOI: 10.1027/2151-2604/a000147

Call for Papers

"Assessment of Competencies in Higher Education"

A Topical Issue of the *Zeitschrift für Psychologie*

Guest Editors: Sigrid Blömeke (Humboldt University of Berlin, Germany), Jan-Eric Gustafsson (University of Göteborg, Sweden), and Richard Shavelson (Stanford University, CA, USA)

The assessment of competence development during the course of higher education presents a substantive and methodological challenge. The challenge is to define and model competence – as the latent cognitive and affective-motivational underpinning of domain-specific performance in varying situations – in a reliable and valid way given the inter- and intra-national diversity of higher education institutions, programs, and teaching processes. Moreover, the question arises as to which criteria may be appropriate to evaluate definitions and measures of competence. Finally, adding to the challenge is the fact that academic programs and labor markets change constantly.

In a globalized knowledge-based society in which demands toward the academic work force become ever more complex, study of higher education's capacity to develop competencies is greatly needed. It is against this background that we invite papers to examine issues of the assessment of competencies, including methodological papers, and papers on the construct definition of competence, including an examination of the reliability, validity, and utility of such assessments in higher education.

The objective of this topical issue is to highlight competence assessment from a substantive and a methodological point of view. The examination of competence measurement structure (e.g., dimensionality or grading) is also of interest. Finally, we seek to cover competence assessment across a wide range of higher education programs and institutions. We welcome full original articles, research notes, and opinion papers. We invite scholars to submit their abstracts on potential papers electronically to the first guest editor and to copy the other guest editors:

Sigrid Blömeke (sigrid.bloemeke@staff.hu-berlin.de)
Jan-Eric Gustafsson (jan-eric.gustafsson@ped.gu.se)
Richard Shavelson (richs@stanford.edu)

How to submit: Interested authors should submit a letter of intent including: (1) a working title for the manuscript, (2) names, affiliations, and contact information for all authors, and (3) an abstract of no more than 500 words describing the content of the proposed manuscript.

There is a *two-stage submissions process*. Initially, authors are requested to submit only abstracts of their proposed papers. Authors of the selected abstracts will then be invited to submit full papers, research notes, or opinion papers. All papers will undergo blind peer review.

Deadline for submission of abstracts is November 15, 2013.
Deadline for submission of full papers is March 31, 2014.

The journal seeks to maintain a short turnaround time, with the final version of the accepted papers being due by August 15, 2014. The topical issue will be published as issue 1 (2015). For additional information, please contact Sigrid Blömeke (sigrid.bloemeke@staff.hu-berlin.de).

General Information About the Journal

The *Zeitschrift für Psychologie* was founded in 1890 and is the second oldest psychology journal in the world. One of the founding editors was Herrmann Ebbinghaus. Since 2007, it is published in English and is devoted to publishing topical issues that provide state-of-the-art reviews of current research in psychology.

For detailed author guidelines, please see the journal's website at www.hogrefe.com/journals/zfp/

DOI: 10.1027/2151-2604/a000148